PHOTON MIGRATION IN TISSUES

PHOTON MIGRATION IN TISSUES

Edited by
Britton Chance
University of Pennsylvania
Philadelphia, Pennsylvania

PLENUM PRESS • NEW YORK AND LONDON

Library of Congress Cataloging-in-Publication Data

Photon migration in tissues / edited by Britton Chance.
    p.   cm.
   Proceedings of a workshop held April 17, 1988, in Philadelphia,
Pa.
  Includes bibliographical references.
  ISBN 0-306-43522-5
   1. Lasers--Physiological effect--Congresses. 2. Laser
spectroscopy--Congresses.   I. Chance, Britton.
  [DNLM: 1. Radiation--congresses. 2. Spectrophotometry-
-congresses.   QD 95 P575 1988]
QP82.2.L3P48 1990
591.19'15--dc20
DNLM/DLC
for Library of Congress   90-7293
                                           CIP

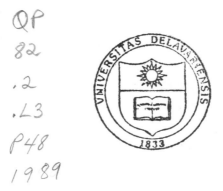

Proceedings of a workshop on Photon Migration in Tissues,
held April 17, 1988, in Philadelphia, Pennsylvania

© 1989 Plenum Press, New York
A Division of Plenum Publishing Corporation
233 Spring Street, New York, N.Y. 10013

Printed in the United States of America

The editor has undertaken to dedicate this volume to the memory of William E. Blumberg, who died June 3, 1989. Bill's tenacious pursuit of the application of physics to new ideas of biology, and his implacable search for the physical basis of biological phenomena clearly distinguishes his work. The breadth of his contribution spanning so many areas of physics is remarkable. Bill took enormous care to explain to his friends and colleagues the physical meaning of biology based on his deep understanding, in order to stimulate their further experimentation. This indeed has been the case in my relationship with Bill. The field and his friends will miss him.

PREFACE

This book is formulated from a number of presentations made at a one-day workshop on the subject of Photon Migration in Tissues. The meeting was held in Philadelphia at the University of Pennsylvania, April, 1988.

The workshop was an impromptu effort to bring together scientists to discuss photon migration in animal tissues and appropriate models.

The rapid emergence of the ideas of Townes and Schalow in their invention of the then called maser, now laser opened up completely unexpected possibilities for biomedical research. Timing of rapid biochemical reaction, identification of unstable intermediates, spectroscopy of short lived fluorescent states were all goals to be expected and achieved. At the same time continuous light spectroscopy of tissue slices and of the myocardium, and eventually of the neonate brain have emerged over the years. Shifting to the red end of the spectrum, Butler and Norris clearly showed how transparent plant materials and the human hand could be illuminated in this region and Jobsis applied their idea to the neonate brain using a multiwavelength technique.

However, the dilemma of these continuous light studies was that photons having traveled any and all path lengths were indiscriminately accepted. This missing information on the optical path length, essential for quantification of tissue spectroscopy, could be afforded by pulsed time measurements, i.e., of the time and distance required for photons to enter the tissue and arrive at the exit point. Thus, the short pulse laser technology provides the pulse timing of the optical path length and therefore solves the dilemma of continuous light spectroscopy of tissues.

A parallel development was a better understanding of the nature of light scattering in tissue and the pathway that photons take in a highly scattering media. Blumberg W.E. (Light propagation in human tissues: The physical origin of the inhomogeneous scattering mechanism. *Biophys. J.* 51:288 (1987)) pointed out that photons of these wavelengths used became completely scattering upon entering a tissue and propagated by a random walk.

The proceedings of this symposium discusses the possibilities of continuous and pulsed light study of tissue models, animal models, and briefly human studies.

The extensive discussion which followed each paper could not be reproduced but the papers themselves speak to the point of the prospects of a new field--Photon Migration in Tissues.

The constituency of speakers was gathered hastily and consequently some important contributions were missed. However, response was enthusiastic, the conference an effective one, and the editor's special thanks are due to each of the authors for their cooperation.

The symposium was made possible by the generous and timely financial support from Becton Dickinson and Company; Coherent (Laser Products Division); Hamamatsu (Photonics Management Corporation); Institute for Structural and Functional Studies, University City Science Center, and 3 M (Medical Surgical Division).

Britton Chance

CONTENTS

Part 3.   INSTRUMENTATION ASPECTS

APPENDIX

SUMMARY REPORT OF A WORKSHOP ON PHOTON MIGRATION

Britton Chance

Department of Biochemistry and Biophysics, University of Pennsylvania
and Director, Institute for Structural and Functional Studies, University
City Science Center, Philadelphia, PA 19104 USA

INTRODUCTION

This workshop was held in Philadelphia on April 17th,1988 was a unique meeting, and perhaps the first meeting of many to follow focusing on a new aspect of spectroscopy in tissues, namely, the migration of photons in tissues, which are poorly resolved under conditions of continuous illumination, but clearly resolved with "time slicing" techniques in the picosecond region. The implications of these techniques for quantitative tissue spectroscopy of absorbing substances in tissue, particularly deoxy-hemoglobin, and for the imaging of hypoxic tissue volumes or tumors is predicted by the theoretical and experimental presentations at the meeting.

THEORETICAL ASPECTS

The theoretical part of the meeting was opened by Chance who traced studies from Tyndall to Keilin and Millikan. W. Blumberg[1] described the theory behind photon migration in tissues. He also emphasized the similarities to migration of photons in tissues to that of diver's visibility in clear and cloudy water or at shallow and large depths. He made clear the generality of the theory and its general applicability to many problems of medicine, physics and optics.

The "modulation transfer function" and its variation with frequency was identified as the thematic parameter of the discussion and many examples of cadaver brains, of skin, and of lakes, etc., were afforded. Particular attention to the distance over which the coherent rays could be observed as a function of optical arrangement, i.e., pinhole optics was a key topic. A direct application to carcinoma and normal tissues indicated that the normal tissue

---

[1] AT&T Bell Laboratories, Murray Hill, NJ

Photon Migration in Tissues
Edited by B. Chance
Plenum Press, New York

propagated more coherent light than did the carcinoma and as the lecture progressed, it then became clear that the range over which the coherent ray was expected to be preserved was very small. Finally, identification of bar patterns by coherent and time resolved spectroscopy were demonstrated with striking results for small range scattering.

In the case of the bar pattern, the variation of the transfer function with radio frequency was best fitted by two Lorenztians, one having to do with short path lengths, and the other with long path lengths. When this analysis was applied to normal and tumor tissues, the long path showed a two-fold decrease for the tumor tissue, and the penetration depth showed a similar decrease over wavelengths from 600 to 1100 nm. Blumberg emphasized the two categories, one well understood where wave functions (Raleigh scattering) and physical optics adequately describe the small scale motions of photons, while large penetration depths wave mechanics are no longer applicable, and statistical processes best describe the photon migration.

Robert Bonner[2] described initial applications of continuous light to platelet absorption where many parameters for the random walk theory of Bonner and Weiss, Nossal et al were given and applications to *ex-vivo* skin were provided at a variety of separations of input and output and hence a variety of path lengths. Generally the absorption coefficient, μ_a, varied from low values of 0.035 cm^{-1} to 0.56. The necropsy tissues were also studied and showed significant differences, for example, atheroma. Finally the question of boundary layers was brought up and the effect of fissures and ventricles in the brain require examination.

James Callis[3] continued the presentation with a description of a color translation system that allowed the detection of hemoglobin absorption in burn tissue which was used as a criterion of viability and rapid healing. A four color system gives a dramatic translation of great utility to the clinician in scoring the viability of the tissue for rapid recovery.

Color translation of infrared absorptions of breast cancer was described where distinctive contrasts with continuous light illumination was obtained providing the tumor, cyst, or other object, was manipulated so that the short path length condition necessary for coherent imaging was feasible. In particular, the pooling of blood in the tumor has provided a high contrast signal. The striking example of coherent ray imaging was provided by a 100 picosecond time slicing device which rejected the long range photons and emphasized the short range ones.

---

[2] Biomedical Engineering Branch, National Institutes of Health, Bethesda, MD
[3] Center for Analytical Chemistry, University of Washington, Seattle, WA

Dr. Brian Wilson[4] next spoke on tissue optical properties in relation to models for light propagation and used suspensions of red blood cells as a basis for a Monte Carlo computation. Patterns of light distribution with deoxy- and oxy-hemoglobin present in the red blood cells were measured at 600 and 665 nm showing how the light propagation was significantly altered by the oxygenation state of the cells, and at the same, time how an absorbing object in the scattering system would generate characteristic patterns of photon migration with continuous light. The term local "light fluence" was introduced and equations for forward and back scatter were presented over wide ranges of wavelengths to indicate Rayleigh scattering, water absorption region, and absorption dominated regions as one progressed from 0.6 to 9 m. *In vitro* muscle and brain models gave absorption coefficients of 0.02 to 0.07 cm^{-1} for muscle, and 0.01 to 0.05 cm^{-1} for brain. Finally, comparison of artificial and real scatters led to a discussion of the nature of the scattering substance itself, which in the discussion was considered to be the cell membrane, but Chance made the point that the scatterers could be membraneous organelles such as the mitochondria and in liver, the endoplasmic reticulum. The relative importance of these scatters and absorbers remains to be determined.

In the discussion chaired by W. Blumberg, diffusion theory and Monte Carlo models were discussed together with their directional dependence. Chance brought up the possibility of picosecond time resolution for a unique solution of the profile of absorption against path length, which is time.

EXPERIMENTAL ASPECTS

R. Greenfeld[5] presented both continuous and pulsed light studies of a milk model, either containing hemoglobin or into which deoxy-hemoglobin could be inserted in the test tube so as to simulate a localized infarct. The effects of the absorber upon the propagation of light was found most clearly exhibited in the logarithmic plot of absorption: log I_0/I vs time or distance. The variation of the logarithmic slope with near and distant absorption of hemoglobin was clearly demonstrated, and compared with results obtained with continuous light. Greenfeld clearly pointed out that the logarithmic intensity function was not linear with distance in the absence of an absorber, whilst it became linear (logarithmic), in the presence of adequate absorption.

Chance described experimental studies of muscle and brain using both continuous and pulsed light, indicating that the time profile of exiting photons was significantly delayed

---

[4] Hamilton Regional Cancer Institute, Hamilton, Ontario, Canada
[5] Department of Electrical Engineering, University of Pennsylvania, Philadelphia, PA

with respect to the 100 picosecond input pulse so that delays of several hundred picoseconds were observed.

He showed that the photon decay curves followed logarithmic functions in skeletal muscle and brain. He further showed that the slope of the logarithmic curve ($\mu_a = 1/L$ log I_0/I) varied significantly with the boundary conditions, being minimal in the largest object studied, human brain and calf; whilst being larger in skeletal muscles and in animal cat head models.

Chance further showed that the presence of deoxy-hemoglobin itself increased the slope (μ) at 760 nm. For a particular case in the cat brain, a maximum for deoxy-hemoglobin absorption in the tissue analysis showed an average deoxy-hemoglobin content of 50 mM. The pulsed light slope increment ($\Delta\mu$) divided by the extinction coefficient, ε, gave 44 mM. The difference spectrum of normoxic and ischemic leg to show a 760 nm peak of deoxy-hemoglobin, as expected, using the pulsed light technique. Experiments in which the hemoglobin concentration is continuously varied were said to be in progress. These experiments validate a theory of photon absorption in which the path length, L, is known and is divided into the absorbancy change (log I_0/I), the quantity, μ_a (dimensions cm^{-1}), is identically equal to eC according to the Beer-Lambert Law. This is a great advantage of the pulse light method, since quantification of hemoglobin concentration changes are often essential. Furthermore, the continuous light method which does not have this intrinsic property, was said to be calibrated by the pulsed light method. The quotient of the observed absorbancy change by continuous light and the value of m gives the effective propagation for the continuous light method, found to be 2 cm for calf muscle, and nearly 10 cm for brain using continuous light geometries developed by Dr. Chance and his colleagues.

M. Tamura[6] , described a proposal for a coherent light imaging system where pulsed light from a laser in the green region would impinge upon the head of the test object. Arrays of detectors would seek the coherent or direct ray and thereby image directly tissue volumes containing hemoglobin in its various states. The estimated power to obtain a coherent response in the adult brain was estimated to be in excess of 20 Watts.

He further described algorithms for continuous light appropriate to the quantification of oxy- and deoxy-hemoglobin in the brain. It was noteworthy that he gave the coefficients of this algorithm.

Finally, studies of milk models and rat heads with pulse light techniques were discussed, raising the question as to whether the boundary conditions for the rat model were

---

[6] Electrotechnical Department, Hokkaido University, Sapporo, Japan

valid for the human head. In fact, much larger absorption coefficients (μ) were observed in the rat head than in the human head, as would be expected according to Chance's theory that photon leakage from the test objects would greatly diminish the migration distance.

Y. Yamashita[7] described the newly developed Hamamatsu NIR 1000 instrument as employing six laser diodes and a photon counting system, showing recent results from the Delpy Laboratory. The system has a capability to detect absorbance changes of 0.02 O.D. on a background attenuation of 10 O.D. with an interval of two seconds.

While the algorithms employed in the neonate brain studies were published in The Lancet, Yamashita indicated that updating of algorithms is possible with their computer system and may indeed have been done to incorporate newer knowledge concerning the real optical path length.

Breast imaging by collimated light showed, for relatively short path lengths and by color translation display, elegant resolution of breast tumors using 632.2 and 830 nm illumination.

Claude Piantadosi[8] described the results of Jobsis' laboratory on cytochrome absorption at 830 nm. They concluded that the oxidized copper absorption in the 830 nm region consists of two peaks, the major one at 822 nm and a minor one at 878 nm. This is in somewhat contradistinction to the low temperature studies of the oxidation of cytochrome aa_3 low temperature with oxygen pulses which showed a component at 740 nm to appear initially, followed by the absorption at the usual position of 830 nm. Why the room temperature and low temperature data appear to be convoluted into different spectral regions is moot. The minor *in vivo* absorber at 878 is subtracted from the major by the algorithms. The Piantadosi algorithms thus involved twelve constants, and data acquisition at four wavelengths, with an apparatus using gallium aluminum arsenide laser diodes as the four light sources. Evaluation of the constants was done in a well tested cat and rat brain models where hemoglobin was nearly completely replaced by fluorocarbon 43. Generally, minimal changes of the output algorithm for cytochrome aa_3 occurred with hemoglobin removal and at the same time, a decrease of oxy-hemoglobin. Whether or not the HbO_2/Hb ratio could be maintained high at the high level of perfusion was not displayed. It was concluded that the maximum error in determination of cytochrome a_3 was approximately 10%, assuming the entire change in cytochrome signal was due to crosstalk from the HbO_2 during the blood washout.

---

[7] Hamamatsu Photonics, K.K., Hamamatsu City, Japan
[8] Department of Medicine, Duke University Medical Center, Durham, NC

Applications of this algorithm to continuous light spectroscopy of skeletal tissues in cardiovascular stress showed parallel changes of cytochrome and hemoglobin + myoglobin over most of the range explored Finally, the anomaly of venous occlusion observed by them to affect cytochrome prior to hemoglobin was pointed to. Whether these changes were dependent upon the transferability of the cat model to the human arm without a correction for boundary condition changes was again moot.

Dr. S. Nioka[9] reported in detail on quantitative studies of multichannel spectroscopy of tissue pigments using a rotating wheel spectrophotometer on the one hand and a rapid scanning Reticon spectrophotometer (Otsuka) on the other from which she derived absolute spectra for cytochrome oxidase *in vitro*. She attempted asynthesis of the *in vitro* data to fit the *in vivo* spectra. Callis reported that the harmonics of water absorption which might well extend into the 800 nm region would have to be taken into account before a meaningful fit was obtained. Detailed parameters of dog and cat models with normal hematocrit and fluorocarbon perfusion were presented by Dr. Nioka with the result that the relationships between redox state of cytochrome, oxygenation state of hemoglobin, and P_aO_2 were obtained in both adult and neonate dogs; the sigmoid hemoglobin characteristics are readily discernable in the data.

Finally, dialogue with Tamura pointed to the important relationships between hemoglobin saturation and energy state of the mitochondria were indicated by their PCr/P_i value verified that the major portion of the response to hypoxia occurs at saturation values of less than 20%, in fair agreement with Nioka. Thus, the region of early warning of incipient hypoxia is between normal hemoglobin saturations of 80 to 90% and 20%. Generally the paper validated the measurement of hemoglobin in brain tissue in neonate and adult animals.

Dr. Ping-Pei Ho[10] represented Dr. Alfano's interest in femtosecond time resolved spectroscopy which was necessarily limited to, or most appropriate to, the measurement of very short distances, in fact, those on a microscopic scale. At an 8 picosecond time resolution, the identification of very small objects, particularly single nails of 1 mm crosssection, was clearly shown for clear solutions; the expected 2 mm resolution was obtained. Such experiments were said to be in the "ballistic" region of light propagation. A system similar to that demonstrated by Callis, but on a shorter time scale, afforded resolution by time slicing of the coherent light emergent from a scattering medium. No doubt the many centers which have picosecond technology can provide highly time resolved time profiles of photon exit from small objects, where femtosecond resolution may be highly

---

9 Department of Physiology, University of Pennsylvania, Philadelphia, PA
10 Department of Electrical Engineering, CUNY, NY

desirable. However, such objects, unlike the human brain, can be dissected to verify the optical ranging geometries.

INSTRUMENTATION ASPECTS (were provided two general papers and a round table discussion)

G. Holtom[11] presented in detail the current status and limitations of resolved system using picosecond laser technology based upon resources available at the University of Pennsylvania's Regional Laser Laboratory. He emphasized the characteristics of Argon Ion and neodymiun Yag (NcYag) lasers, balancing stability of the source--the argon ion lasers against the greater reliability of the Yag lasers of the Antares configuration. Consideration of the detectors is paramount, particularly avoidance of pile-up problems in single photon detection. This system, together with the similar system at Palo Alto, were the ones on which Chance's experimental feasibility criteria have been established, namely, a 100 picosecond input pulse to the adult human brain giving adequate time resolutions of photon migration over the approximately 1-2 nsec interval described above. Thus a 12 m channel plate has proved itself to be adequate for these studies Great advantages in the acquisition of data appropriate for imaging algorithms is anticipated if the 16 anode version of the microchannel plate tube becomes available.

Dr. Joe Lakowicz[12] : Phase modulation of the CW light by an acousto-optical modulator had a frequency limitation of 250 MHz whilst 50 psec pulse trains have modulation useful in the GHz region from which the recovery of the phase shift was 100% even though a particular harmonic (100-1000 MHz) was injected into the phase demodulator system. The system limitation of the detectors for the squirrel cage R928 tube appeared to be 330 MHz, the 2-inch end-on tube, 600 MHz (R1828) while the micromulti-channel plate tube was of an order of magnitude faster, 6 GHz. In short, the time capabilities of phase demodulation appears as good as that using direct pulse timing; the characteristics of the detector being limiting in both cases. Values of 30-50 ps appeared to be characteristic of the technology available in the University of Pennsylvania Regional Resource Laser Lab (Holtom) and Lakowicz's lab in Baltimore.

The Round Table Discussants[13-17] put forth two important possibilities, one by Lakowicz that the techniques employed for 2D X-ray imaging may well become useful in

---

11 Regional Laser Laboratory, University of Pennsylvania, Philadelphia, PA

12 Department of Biological Chemistry, University of Maryland, Baltimore, MD

13.L. Brand, Johns Hopkins University, Department of Biology, Baltimore, MD

14.J. Knutson, Laboratory of Technical Development, National Institutes of Health, Bethesda, MD

15. H. Oka, Otsuka Electronics, Osaka, Japan

optical studies whilst Callis pointed out that frequency chirping of laser diodes might render them useful in the phase modulation domain.

SUMMARY

Those present at the meeting agreed that it was a most timely and important gathering of new ideas on a new topic and that the future for optical ranging and imaging of highly scattering media in which photon migration represented the propagating algorithm was promising. The method of time resolution of coherent directly transmitted radiation was possibly limited to smaller objects and simpler systems than the adult human brain. The detection of deoxy-hemoglobin, localized or globally, in the human brain appeared to be most useful, not only for the identification of the presence of the deoxy-hemoglobin but especially to assist in the interpretation of Magnetic Resonance Imaging (MRI) data, the contrast of which for both T_1 and T_2 depends critically upon knowledge of whether oxy-hemoglobin, deoxy-hemoglobin or met-hemoglobin is present.

[13.]L. Brand, Johns Hopkins University, Department of Biology, Baltimore, MD

[14.]J. Knutson, Laboratory of Technical Development, National Institutes of Health, Bethesda, MD

[15.] H. Oka, Otsuka Electronics, Osaka, Japan

[16.] A. Predham, Department of Electrical Engineering, CUNY, NY

[17.] J. Vanderkooi, Department of Biophysics and Biochemistry, University of Pennsylvania, Philadelphia, PA

Part 1. Theoretical Aspects

A RANDOM WALK THEORY OF TIME-RESOLVED OPTICAL ABSORPTION SPECTROSCOPY IN TISSUE

Robert F. Bonner[a], Ralph Nossal[b] and George H. Weiss[b]

[a]Biomedical Engineering and Instrumentation Branch
[b]Physical Sciences Laboratory
National Institutes of Health
Bethesda, MD 20892

ABSTRACT

A random walk theory of photon diffusion in tissue is discussed, and shown to reproduce features of some recent experiments by Chance et al. The theory allows one to calculate, from the absorption and scattering coefficients of turbid medium, the time dependence of surface emission.

INTRODUCTION

Optical remote sensing of living tissues and fluids is an attractive way to continuously monitor important physiological parameters such as tissue oxygenation, metabolic activity, membrane potential, and blood flow. A variety of prototype and commercial systems currently are used to do such tasks. Quantitative optical signals are critically dependent on optical path length, yet any detected signal from tissue derives from a wide distribution of optical paths which sample different volume elements. The distribution of these trajectories depends on the scattering and absorption properties of the tissue as well as the external probe geometry. Until now, neither theory nor measurements of light migration in tissues have provided accurate quantitation of these trajectories. Consequently, remote-sensing optical instrumentation usually has been empirically calibrated against measurements involving techniques which are themselves difficult to implement and may not be exactly equivalent.

To provide a theoretical basis for such measurements, we have applied random walk theory to the problem of light migration in turbid media[1]. The analysis is specifically

Photon Migration in Tissues
Edited by B. Chance
Plenum Press, New York

directed towards improving quantitative interpretation of optical measurements made noninvasively at the surface of a tissue. This work has had two objectives: 1) to use remote measurements at the surface to determine optical properties of the bulk tissue and 2) to use these optical properties to determine distributions of photon path lengths and photon sampling efficiencies. These are needed for quantitative interpretation of optical remote sensing of physiological parameters. For example, we previously developed a laser-Doppler instrument to noninvasively monitor microcirculatory blood flow[2]. This system injects laser light at a point on the surface and collects it at another point on this surface, using small optical fibers. The detected light which has migrated in the tissue between these two points is processed in real-time to give the mean red blood cell speed from the mean Doppler broadening and the fraction of the light that is Doppler-shifted (which increases with both the optical path length and the concentration of moving red blood cells in the tissue). Without knowing the details of the photon paths, we developed a theoretical model allowing quantitative use of this technique. However, because one lacks knowledge of the distribution of path lengths of detected photons, calibration of this technique requires comparison with other, independent, measurements of microcirculatory blood flow. Similarly, without appropriate theory we cannot infer how different tissue elements in the vicinity of our probe contribute to the signal. The same problems exist with measurements of tissue oxygenation based on changes in optical absorption of hemoglobin and myoglobin.

Considerable interest recently has been directed to time-resolved *in vivo* spectroscopy, particularly the measurement of the probability distribution of times for photons to travel from one point on a tissue surface to another (Chance and coworkers[3,4]). The time course of photon escape from another point on the surface a few centimeters away has been measured by using picosecond laser pulses directed via optical fibers onto a tissue. Since the transit time is the optical path length divided by the speed of light, the attenuation of the signal with time must in part depend on the average absorption within the tissue through which the light has migrated[3,4]. The attenuation of the signal due to absorption is simply the product of the molar extinction coefficient, the pathlength, and the concentration of the absorbing species. Since the first two factors can be determined, these measurements might provide absolute values of the concentration of the absorbing molecules[3,4]. There are, however, two complications: 1) the distribution of the transit times also depends on the tissue scattering properties and the measurement geometry (a complex function even in the absence of any absorption) and 2) emergent photons may sample various regions of the tissue, which contain different concentrations of absorbing species. For example, the noninvasive transit-time measurements of oxygenation in brain[3] and muscle[4] seem to depend on measurement geometry (which cause variations in the relative probabilities of different optical paths). Resulting parameters must, in some way, indicate the oxygenation of the scalp/skull as well as that of some ill-determined part of the underlying brain (or of the skin and the underlying muscle). Therefore, in order to interpret the empirical transit time data, it is important to ascertain the probability distribution of optical paths for different measurement geometries, taking into account both absolute tissue absorption and the relative sampling of different depths. We previously applied random walk theory to derive expressions for many of the relevant distributions[1].

Tissues are such strong diffusers of light that the mean free path between anisotropic scattering is on the order of 50μm, and the light is completely randomized within ~1 mm. Thus, if the direct path from the two points on the tissue surface is much greater than 1mm, the migration path of the photons may be adequately described by a random walk of many steps. In the case of the measurements of Chance et al., the mean transit times between points which are separated by ~40 mm are on the order of 1 nsec where the speed of light is ~0.23 mm/psec, so the mean path lengths are of the order of 230 mm (n ~ 200 isotropic step pathlengths, L). This suggests that we can analyze these highly randomized photon paths in terms of a theory in which the details of anisotropic scattering are irrelevant and for which the approximation of a large number of steps n is appropriate.

DESCRIPTION OF THE MODEL

Our model approximates the motion of a photon in a tissue as a random walk on a semi-infinite simple cubic lattice. The type of lattice can be shown to be unimportant for later developments. The coordinate system is indicated in Figure 1, where the direction of the incident beam is taken as the positive z axis and the surface lies in the x-y plane. The tissue surface is assumed to be infinite in the x and y directions and in the simplest version infinite in the z direction as well. As we shall see later, few photons detected at the surface have in fact penetrated to significant depths, so this requirement is not restrictive. Some further physical assumptions used in our analysis are:

1. Absorption within the tissue is governed by Beer's law.
2. Photons reaching the surface contribute to the light intensity measured there.
3. Material properties of the tissue are isotropic.

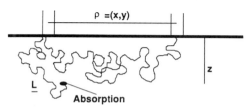

Fig. 1.   A schematic of light migration between two points on the tissue surface (in the x-y plane) separated by $\rho = (x^2+y^2)^{1/2}$. The photon direction is randomized over a distance L << ρ. Two photon paths are shown: one reaching the surface at r and detected as surface emission, the other absorbed at depth z.

The theoretical development is based on a lattice random walk which requires that two parameters be furnished. The first is L, the lattice spacing (which is the reciprocal of the rms effective cross section for isotropic scattering); and the second is μ, the absolute absorption coefficient α times L, the path length for one lattice step (μ=αL). These two parameters, μ and L, generally are found by curve fitting to empirical data. In the subsequent analysis, the lattice coordinates denoted by (x,y,z) are dimensionless integers and correspond to the physical coordinates $(x,y,z)=(xL,yL,zL)$.

In our model, the random walk takes place in discrete time (i.e., the photons take steps of length L at times n=1,2,3...), but terminates when the photon is either absorbed within the tissue or is re-emitted on reaching the surface at z=0. At step n=0 the photons are located at (0,0,1). The step number, n, can be related to the physical time-of-flight, t, (and path length l) by:

$$n = c_T t/L = l/L \qquad (1)$$

c_T being the speed of light in tissue (~0.23 mm/psec). We assume that the photon travels at uniform speed or, equivalently, that the macroscopic index of refraction does not vary significantly within the tissue.

Under our sumption of isotropic scattering, the random walk proceeds as steps to one of six nearest neighboring lattice points, each chosen with equal probability. This may appear unduly restrictive, as physical scattering within tissue is highly anisotropic. However, the model seems to describe light migration for all but very small distances, and produces results very close to Monte Carlo simulations of a more realistic anisotropic process when an appropriate equivalent rms isotropic scattering length is defined[5]. Theoretical considerations deriving from the central limit theorem imply that an analysis based on equivalent isotropic scattering leads, for long paths, to the same results as those for anisotropic scattering. Our simple model does lead to instructive analytical results, and we will leave a fuller discussion of anisotropy to a subsequent paper. To parameterize internal absorption in the random walk model, we assume that the probability of a random walker making a single step without being absorbed is equal to exp(-μ) or, equivalently, $exp(-\alpha L)$.

The probability that a walker is at site r=(x,y,z) at step n in the presence of a nonreflective boundary at z=0 will be denoted by $Q_n(r)$. This, in turn, can be written in terms of the analogous probabilities, $P_n(r)$, for a random walker moving without absorption on a lattice infinite in all directions as

$$Q_n(r)= [P_n(x,y,z-1)-P_n(x,y,z+1)]e^{-n\mu} \qquad (2)$$

which satisfies the boundary condition $Q_n(x,y,0)=0$. This is equivalent to the statement that photons which reach the tissue surface vanish from the system and constitute the surface intensity. The reason this approach is advantageous is that we can write an explicit expression for $P_n(\mathbf{r})$. However, we will use a somewhat simpler result which is valid for large n $(n \geq 10)$;

$$P_n(\mathbf{r}) \sim \{3/(2\pi n)\}^{3/2} \quad \exp\,[-3(x^2+y^2+z^2)/(2n)] \qquad (3)$$

Equations (2) and (3) allow us to find the probability distribution needed for the interpretation of transit time data and remote sensing measurements which depend on path length. This function will be denoted as $\Gamma(n,\rho)$, which is the probability that a photon reaches the surface at a distance ρ from the injection point at step n, where $\rho=(x^2+y^2)^{1/2}$. Since time-of-flight measurements typically consist of data accumulated over many laser pulses at a given ρ, the absolute probability may not be determined. The conditional distribution of transit times at a given ρ is given by $\Gamma(n|\rho)$, which is the probability that a photon reaches the surface at z=0 at step n given that it is detected at ρ. Since, according to Eq. (1), n is proportional to the time-of-flight, t, these functions give the probability distribution of time-of-flight between two points on a tissue surface.

The joint probability distribution of n and ρ can be written in terms of $Q_n(\mathbf{r})$ as

$$\Gamma(n,\rho) = (1/6)\ Q_{n-1}(x,y,1)\ e^{-\mu} \qquad (4)$$

since the photon, in order to reach the surface z=0 at step n, must be at z=1 at step n-1 and then move with probability 1/6 to z=0 without being absorbed. On substituting Eq. (3) into Eq. (2) and then into Eq. (4), we find that

$$\Gamma(n,\rho) = [3/(32\pi^3 n^3)]^{1/2}\ (1-e^{-6/n})\ \exp[-3\rho^2/(2n)-\mu n] \qquad (5)$$

The conditional probability $\Gamma(n|\rho)$ is determined from the joint probability $\Gamma(n,\rho)$ by

$$\Gamma(n|\rho) = \Gamma(n,\rho)\ /\sum_{n=0}^{\infty} \Gamma(n,\rho). \qquad (6)$$

However, since a large n approximation has already implicitly been used in writing the expression for $P_n(\mathbf{r})$ in Eq. (3), we can replace the sum in the denominator by

$$\Gamma(\rho) \sim \int_0^{\infty} \Gamma(n,\rho)dn. \qquad (7)$$

Thus, from Eq. (5), $\Gamma(n|\rho)$ is to a good approximation given by

$$\Gamma(n|\rho) \sim \{\rho/g(\rho)\} \quad [3/(2\pi n3)]^{1/2} \quad (1-e^{-6/n}) \quad \exp(-3\rho^2/2n-\mu n) \tag{8}$$

in which $g(\rho)$ is the function

$$g(\rho) = \exp\{-\rho(6\mu)^{1/2}\} - [\rho/(\rho^2+4)^{1/2}] \quad \exp\{-[6\mu(\rho^2+4)]^{1/2}\} \tag{9}$$

For each value of ρ, the distribution of times-of-flight given Eq. (8) shows a rapid rise to a peak value (which occurs at a time much greater than the straight-path transit time, r'/c_T), followed by an approximately exponential decay (Figure 2), which approaches $e^{-\mu n}$ (or $e^{-\alpha l}$) as n approaches infinity. Thus, if sufficient signal-to-noise permits photon detection at large times, a reasonable estimate of the average absolute tissue absorption coefficient, α, can be obtained, with corrections due to the additional dependences on n in Eq. (5). One observes in Figure 2 that the detected intensity, at any given point ρ, has a maximum at n_{max} (where $n_{max} = t_{max}(c_T/L) = l_{max}/L$), which is approximately at the value of n for which the exponential factor in Eq. (8) is a maximum,

$$n_{max}(\rho) = \rho[3/(2\mu)]^{1/2}. \tag{10}$$

Previously we showed that the mean path length at a given value of ρ, $<n|\rho>$, is approximately equal to this value (see Eq. (17) of ref. 1), even for smaller ρ. This relationship allows an estimate of both μ and L, once the absorption coefficient, α, is determined from the decay of photon transit times.

These tissue optical constants are representative of the region of the tissue traversed by the detected photons. We recently calculated statistical properties of the depth probed by photons emerging at the surface of the tissue, given that a photon is emitted after n steps (or at time $t = nL/c_T$). These are expressed in terms of, $p(z|n)$, a probability density of residence times at a depth z given by[5]

$$p(z|n) = (12z/n) \quad \exp\{-6z^2/n\} \tag{11}$$

Thus the depth at which the photon has spent the maximum amount of time is

$$z_m = \{n/12\}^{1/2} \tag{12}$$

and the average depth calculated from Eq. (11) is $<z> = (\pi n/6)^{1/2}$, so that both z_m and $<z>$ are proportional to $n^{1/2}$. In these expressions, there is no dependence on μ or ρ. The lack of dependence on μ is readily understood, if we note that only those photons which emerge at a specific time are significant, even though the total number that emerge decreases with increasing μ. The lack of ρ dependence is more subtle. It occurs because our expression for

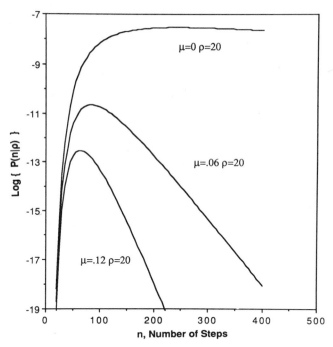

Fig. 2. The predicted $\log_{10}$ of the probability of photons being emitted from a surface at point $\rho=20$ after n steps for three different absorption coefficients ($\mu=0$, .06 and .12). This probability $P(n|\rho)$ is equivalent to that of light being emitted at a point r' ($=\rho/L$) cm from the source at a time t ($=nL/c_T$).

representation for $P_n(r)$ of Eq. (3) does not decay too greatly from its maximum, thereby excluding the tails of the function (i.e., large ρ at fixed n). In the tails of the distribution of Eq. (11), there is indeed a dependence of z_m on ρ, but the calculation of the dependence is quite complicated.

QUALITATIVE RESULTS

In Figure 2, three curves predicted by Eq. (5) are shown for the path length or time dependence of the decay of surface intensity at a point ($\rho=20$) for different absorption coefficients ($\mu=.00,.03,.06$) following the instantaneous light input at t=0. The characteristic features of empirical curves reported by Chance[3,4] are clearly seen: a delayed rapid rise to a peak intensity which occurs at t_{max} much larger than the straight-path transit time, followed by a decay which exhibits a rate dependent on μ. At large n (or time), the apparent absorption coefficient determined from this exponential decay rate is an accurate estimate of the actual μ or α. Here the relative light intensity $\Gamma(n|\rho)$ varies as $n^{-5/2} \exp\{-3\rho^2/(2n)\} \exp(-\mu n)$, for which the last factor dominates (for $\mu>0$ and large n).

17

The result of fitting only the first 3-4 decades of the decay for curves in Figures 2 and 3 (as Chance has done[3,4]), leads to values of the absolute tissue absorption that are less than the actual values. The factors that are independent of absorption in Eqs. (5) and (8) are responsible for these deviations. They describe the time dependence of surface intensity in the absence of any tissue absorption. In the case of no absorption ($\mu=0$ in Figure 2), the intensity rises to a maximum only after a significantly longer delay and is followed by a much slower decay. The rising phase is due to an increasing number of possible paths between the two points as the total path length increases, whereas the subsequent decay is due to the depletion of photons within the tissue by re-emission from the surface. When absorption occurs ($\mu \neq 0$), the slow rising phase of the non-absorptive factors, $n^{-5/2}$ $\exp\{-3\rho^2/(2n)\}$, reduces the rate of decay immediately following n_{max}. Therefore, errors in estimates of absolute absorption become less at longer times and smaller ρ (fiber separation). Unfortunately, at longer times the signal is highly attenuated and the resulting poor signal-to-noise requires averaging an extremely large number of laser pulses to get an accurate estimate of μ.

In Figure 3 we examine the effect of changing the tissue scattering length L, while maintaining the absolute absorption α ($=\mu/L$) constant and measuring at the same surface point r'($=\rho L$). At this point, more light is detected as ρ decreases (i.e., the scattering length L increases), and the decay reaches its asymptotic limit more quickly. As indicated in Eq. 10, the time of maximal intensity varies strongly with the scattering length. Thus the time to peak value might provide additional information on the tissue optical properties. Rather than relying on the decay rate at long times, it would appear better to use the theoretical equations to calculate the absorption and scattering coefficients of the tissue (as discussed in the previous section) from the temporal rise and fall in the region with the best signal-to-noise.

Chance's initial results attempt to monitor changes in hemoglobin oxygen saturation within tissue. In this case the change in absorption coefficient is known, and the tissue scattering properties presumably do not change significantly during an experiment. The ratio of two curves predicted by Eq. (5) for conditions which are identical except for a variation in absorption from μ_1 to μ_2 is simply an exponential decay with time (i.e., $\exp\{-(\mu_1-\mu_2)n\}$). By taking the ratio of two curves of intensity vs. time following laser pulses onto a given tissue with varying oxygenation, one should get an exponential decay dependent only on the change of absorption. This method would have the empirical advantage of using the portion of the curve with the best signal-to-noise and requiring fewer photons (less energetic laser pulses or fewer number of averaged pulses). It is limited, however, to measuring changes in the absolute absorption and not the absolute values themselves.

In these time-resolved measurements at a fixed ρ, the depth sampled increases with $t^{1/2}$. For a typical measurement on tissue (L=1mm, ρ=50, r'=50mm, n=200-400), the mean sampled depth, <z>, is approximately 10-14 mm. Although this mean depth does not

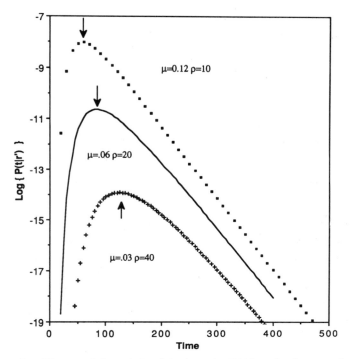

Fig. 3. The predicted $\log_{10}$ of the relative light intensity $P(t|r')$ emitted at a point r' cm from the source after a time t for three different ρ values (and scattering lengths, $L=r'/\rho$) and a constant absolute tissue absorption coefficient ($\alpha=\mu L$). Note that all these curves have the same slope at large times. The arrows indicate t_{max}.

vary with tissue absorption, increasing μ will decrease the signal at large n, making measurements there more difficult. Thus high tissue absorption will probably limit analysis to short times and more shallow depths. Furthermore, as indicated in Figure 3, it might be advantageous to measure the decay rates at small ρ (so that $\rho \ll (2\mu/3)^{1/2}n$ at smaller n). This, however, would result in a more superficial volume being sampled. If the tissue optical properties change with depth, the increase in sampling depth with increasing time could result in deviations from simple exponential decay. This might provide a method to look for such optical discontinuities.

LAYERED MEDIA

Chance et al. have illustrated the possibility of using pulsed light spectrophotometry for non-invasive examination of oxygen delivery to the brain[3] and muscle[4]. The circumstances under which the technique would be applied differ somewhat from the situation described above, as the medium which the light penetrates might be quite inhomogeneous. Photons must pass through bone (the skull) as they enter and emerge from the head, thereby affecting the delay times for reemission[6]. Additional complexity might arise from local heterogeneity in the absorption coefficient related to different degrees of tissue oxygenation. How might such optical features affect interpretation of the shape of the emergent photon pulses?

We recently examined several aspects of photon migration in models of layered tissue[6]. Unfortunately, the analytical theory for heterogeneous structures is much more complicated than that for homogeneous media. Hence we devised an iterative numerical scheme for computing lattice occupancy probabilities, and applied it to a study of migration within a structure composed of a bulk medium clad with a superficial layer having a differing absorption coefficient (see Figure 5). Although the model calculations were performed for planar geometries (i.e., flat surfaces), they should yield insight, also, into the behavior of photons migrating within structures having only a modest degree of curvature, such as certain regions of the head.

The results can be summarized qualitatively as follows. Let us consider, first, the case where the optical absorbance coefficient of the surface layer (μ_1) is much greater than that of the underlying bulk material (μ_2). Photons which are detected close to the point of insertion (small ρ in Figure 4) mostly will have moved through the top layer before reaching the detector. However, photons detected far from the insertion point mostly will have moved through the lower layer. This is well illustrated in Figure 5, where we show <n|r>, the expected path length traversed by photons which emerge within a ring located a distance ρ lattice units away from the point of insertion, for the case $\mu_1/\mu_2 = 0.2/.01 = 20$ (see Ref. 6 for details). The numbers besides the various curves correspond to T, the thickness of the top layer in lattice units. Consequently, T=0 corresponds to a homogeneous sample having an absorption coefficient μ_2 (i.e., that of the bulk), and T=∞ corresponds to a homogeneous sample whose absorptive properties are those of the top layer

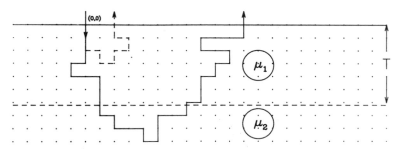

Fig. 4. Photon migration in layered media. Top layer, which is T units thick, is characterized by absorption μ_1. Bottom layer (infinite in extent) is characterized by a different absorption coefficient, μ_2 (see Ref. 6).

(μ_1). The scattering lengths here are considered to be identical in both regions. In this case the function <n|r> shows a rather abrupt transition, at a distance r* that increases as the layer thickness increases.

The situation relating to measurements of blood oxygenation within the head probably is at the other extreme. The highly perfused brain tissue may absorb infrared radiation to a higher degree than does the skull, so that $\mu_1 < \mu_2$. In such cases <n|r> does not show a clear transition from the behavior of the upper layer to that of the lower[6], but the presence of the surface layer does have a great effect on the average transit time. Somewhat exaggerated behavior is shown in Figure 6 (for μ_1/μ_2 = .01/0.2 = .05), but qualitatively similar behavior is found when μ_1 is taken to be close in value to μ_2 .

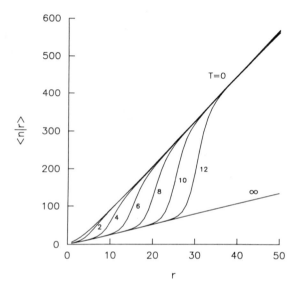

Fig. 5. Expected path length <n|ρ> for photons emerging at distance ρ. Numbers beside curves pertain to the thickness of the upper layer. (μ_1=0.2, μ_2=.01).

These computations clearly show that the average transit time is affected by the presence of a surface layer. We plan to extend our study to obtain distributions of transit times in layered media as well as the mean transit times (or pathlengths) that we previously obtained. We then will be able to investigate how heterogeneity affects the various detectable features of emitted pulses. Chance et al.[3,4] have reported, using a cathead model, that removal of the skull from the brain indeed does change the decay rate of the long-time portion of the detected signals.

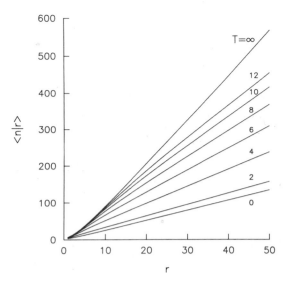

Fig. 6. Expected path length when absorption of the surfacelayer is less than that of the underlying matrix, $\mu_1=.01$, $\mu_2=0.2$ (cf. Fig. 5).

CONCLUSION

Photon transit time distributions following a brief laser pulse provide a new way to look at attenuation with path length. The tissue scattering and probing geometry have important effects on the distributions obtained empirically. A random walk theory describes these effects and suggests a variety of data analysis schemes. Such analysis of this time-resolved data would appear to allow the noninvasive measurement of both the absolute absorption coefficient and a characteristic scattering coefficient of the tissue. The effects of nonuniform media are complex, but time-resolved data may assist in detecting inclusions in tissues.

REFERENCES

1. R. F. Bonner, R. Nossal, S. Havlin and G. H. Weiss, Model for photon migration in turbid biological media, *J. Opt. Soc. Am.*, A 4:423-432 (1987).
2. R. F. Bonner, T. R. Clem, P.D. Bowen, and R. L. Bowman, Laser-Doppler real-time monitor of pulsatile and mean blood flow in tissue microcirculation, in: "Scattering Techniques Applied to Supramolecular and Nonequilibrium Systems," S. H. Chen, B. Chu and R. Nossal, eds., Plenum, New York (1981).
3. B. Chance, J. S. Leigh, H. Miyake, et al., Comparison of continuous and pulsed light measurements of deoxyhemoglobin in Brain. *Proc. Nat. Acad. Sci. USA*, 85:4971-4975 (1988).

4. B. Chance, S. Nioka, J. Kent, K. McCully, M. Fountain, R. Greenfeld and G. Holtom, Time resolved spectroscopy of hemoglobin and myoglobin in resting and ischemic muscle, *Anal.Biochem,* 174:698 (1988).

5. G. H. Weiss, R. Nossal and R. F. Bonner, Statistics of penetration depth of photons re-emitted from irradiated tissue, *J. Modern Optics,* 9 (1989).

6. R. Nossal, J. Kiefer, G. H. Weiss, R. Bonner, H. Taitelbaum and S. R. Havlin, Photon migration in layered media, *Applied Optics,* 27:3382-3391 (1988).

TISSUE OPTICAL PROPERTIES IN RELATION TO LIGHT PROPAGATION MODELS AND IN VIVO DOSIMETRY

Brian C. Wilson, Michael S. Patterson, Stephen T. Flock, and Douglas R. Wyman

Hamilton Regional Cancer Centre
and McMaster University
711 Concession Street
Hamilton, Ontario, Canada L8V 1C3

INTRODUCTION

Optical radiation (ultraviolet, visible, infrared) propagates through tissue absorption and scattering processes take place which depend strongly on the wavelength of the radiation and which may vary considerably between tissues. With the widespread and increasingly varied applications of light in medicine and biology, there is a need for accurate but practicalble methods to determine the spatial (and possibly also the temporal) distribution of light in tissue. There are two approaches to the problem.

1. Optical fibers, coupled to a photodetector, may be inserted into tissue during light irradiation in order to measure directly the local photon or energy fluence rate. With a simple cut-end fiber, the limited numerical aperture of the detector provides a directional dependence that can be used to determine the angular radiance pattern at depth (see Figure 1A). However, in a region where the radiance is highly anisotropic (e.g. near the surface of irradiated tissue), multiple insertions of the fiber to the same point in tissue at different angles are required, as illustrated in Figure 1B, so that the true fluence rate can be obtained by integration over solid angle. This type of interstitial probe has been applied by a number of investigators, both in vivo and in .vitro[1-4]. Recently, optical fibers with a small scattering sphere at the tip have become available which have an approximately isotropic response so that the fluence rate at a point can be determined by a single measurement, as shown in Figure 1C. Such detectors must be calibrated to correct for residual response anisotropy, and for refractive index mismatch between the scattering tip and the tissue[5,6]. The main drawback to using optical fiber detectors to measure the fluence distribution

within tissue is their invasiveness. In vivo, and particularly in patients, this limits the number and locations of measurements.

2. The second approach is to calculate the spatial distribution from a knowledge of the irradiation geometry, the tissue geometry and the optical properties of the tissue. This requires appropriate models to describe the propagation of light, which ideally should be able to accept the complex geometries found in many applications and the range of optical properties characteristic of living tissues. Patterson et al.[7] have recently reviewed the models which have been proposed to date.

Formally, the correct theoretical description of light propagation in tissue would involve solving Maxwell's Equations with the appropriate boundary conditions. Ishimaru[8] has described a method based on multiple scattering electromagnetic theory in which the tissue is specified as a medium of statistically varying electrical permittivity. However, such methods have not lead yet to practical models for use in realistic biological or clinical situations.

The alternative, radiative transfer theory, is less rigorous in terms of the real physics of light propagation, but nevertheless has lead to useful workable models for practical problems. In essence, the wave properties of the radiation are ignored, and the photons are regarded as neutral particles propagating through a medium which can be described by its absorption, scattering, and differential scattering coefficients. Wave phenomena such as interference and diffraction, if they occur in tissue, cannot be taken into account. A further caveat is that care must be taken in the physical interpretation of the optical coefficients, especially the scattering values. Thus, for example, many soft tissues have differential scattering distributions ("phase functions") in the visible spectrum which are strongly forward peaked[9], implying that the scattering is Mie-like, i.e. as in the scattering from discrete particles large in diameter compared with the wavelength of light, suspended in a medium of different refractive index. However, it may

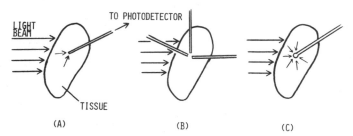

Fig. 1. Illustration of measurements of A) the total diffuse reflectance, R, from an irradiated tissue, and B) the local diffuse reflectance, R(ρ), as a function of radial distance ρ from an incident pencil beam.

not be correct to ascribe a corresponding physical scattering particle size to tissue and to conclude that such particles "cause" the scattering in tissue, which is, rather, a continuous optical medium of spatially varying refractive index.

Accepting these restrictions, radiation transfer theory in an optically homogeneous medium leads to the Boltzmann equation.

$$\hat{\Omega} \cdot \nabla L (\underline{r}, \hat{\Omega}) + \mu_t (\underline{r}) L (\underline{r}, \hat{\Omega}) = \int_{4\pi} d\hat{\Omega}' \frac{d\mu_s}{d\Omega} (\underline{r}, \hat{\Omega}' \rightarrow \hat{\Omega}) L (\underline{r}, \hat{\Omega}') + S (\underline{r}, \hat{\Omega})$$

(1)

where $L (\underline{r}, \hat{\Omega})$ is the energy radiance in direction $\hat{\Omega}$ at position $\underline{r}$ in the medium, S is the photon source, and the total and scattering coefficients are μ_t and μ_s, respectively ($\mu_t = \mu_s + \mu_a$, where μ_a = absorption coefficient). The derivation of this equation may be found, for example, in Duderstadt and Hamilton[10]. There is no general analytic solution to equation 1, although methods for specific simple geometries have been developed (see ref. 7). There are a number of approximate or practical methods, two of which, diffusion theory and Monte Carlo simulation, will be discussed in detail below.

Even for simple irradiation and tissue geometries, the choice of model is strongly dependent on the tissue optical properties, particularly on the tissue albedo ($a = \mu_s / (\mu_a + \mu_s)$) and the scattering anisotropy (g = average cosine of the scattering angle). Three regimes may be identified:

1. absorption dominated,
2. scatter dominated,
3. comparable scatter and absorption.

In the following section, the wavelength ranges corresponding to these different cases will be identified, and the current knowledge of the fundamental optical properties of tissues will be summarized. We will then discuss the advantages and limitations of diffusion theory and Monte Carlo models in handling light propagation calculations for the different albedo and anisotropy conditions. Finally, we will consider potential future methods for determining the optical properties of tissues non-invasively in vivo, which will be essential in applying the models to calculate the spatial distribution of fluence in specific clinical cases.

TISSUE OPTICAL PROPERTIES

Figure 2 shows the absorption spectra for water, hemoglobin and melanin throughout the UV, visible and IR range. These are the major absorbing components of many soft tissues. This graph shows the approximate location of the three

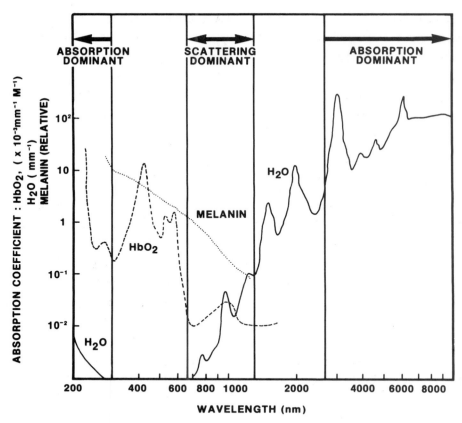

Fig. 2. The absorption spectra of water, oxyhemoglobin and melanin (after ref. 11).

absorption/scatter regimes, applying the somewhat arbitrary criterion that $\mu_a > 10\ \mu_s\,(1-g)$ for the absorption dominated case and $\mu_s(1-g) > 10\ \mu_a$ for the scatter dominated case. These limits correspond roughly to where Beer's Law and diffusion theory apply, respectively, as discussed below. Absorption is dominant in the UV at wavelengths below about 300 nm and in the far IR. The scattering dominated regime holds only in the so-called "therapeutic window" in the far visible and near IR (~600-1200 nm). There are two regions where scatter and absorption are comparable: UVA and most of the visible, and an IR band from about 1300-2500 nm.

Measurements of tissue optical properties are rather sparse. Most studies have been either in skin, particularly in the UV, or at around 630 nm where the clinical use of photodynamic therapy has provided strong impetus to develop accurate light dosimetry. As described by Wilson et al.[12], the techniques to measure the fundamental optical absorption and scattering properties can be split into two classes:

a. _direct_ measurements in which optically thin tissue samples are used (i.e. thickness << $1/\mu_t$). This allows direct determination of μ_t (by a narrow-beam attenuation experiment), of the scattering phase function (by goniometry) and, possibly of μ_a (by using an integrating sphere containing the irradiated tissue sample to detect all the scattered light). These techniques may, in principle, be applied to any of the three scatter/absorption regimes, although great care must be taken in the scatter dominated case to achieve truly optically thin samples, i.e. those in which multiple scattering is negligible[9,13].

b. _indirect_ measurements of some macroscopic, or "bulk", tissue optical parameters, such as diffuse reflectance and transmittance[13,14], or the distribution of energy fluence and/or the pattern of radiance as a function of depth in tissue below the irradiated surface[1,2,15], from which the fundamental absorption and scattering properties can be derived by applying specific models of light propagation.

Absorption dominated

The main data in this case are for the absorption coefficient, which in the far-IR is essentially the same as for water. In the UV there are additional contributions to absorption from bio-molecules such as nucleic acids and proteins. In both regions the absorption coefficient may be extremely large, so that the fluence falls off very rapidly with depth. For example[16], at the greatest water absorption peak at around 2900 nm, $\mu_a > 100$ mm^{-1}. Thus the fluence is reduced to 1% of the incident value at a depth of only 50 μm below the irradiated tissue surface.

Scatter dominated

The data at 630 nm, which lie at the lower end of the scatter dominated range, show the following "generic" characteristics for a variety of soft tissues: μ_t ~10-100 mm^{-1}, a>0.99, g>0.6. These characteristics have been determined by both direct[9,13] and indirect[2,15] techniques. Although the total attenuation coefficient is large, the very high albedo combines with the forward scattering to give substantial penetration of light into tissue. Thus, for example, the effective penetration depth of 630 nm light in tissues (defined as the 1/e depth) is in the approximate range 1 to 10 mm, depending on the blood content and pigmentation (see ref. 17).

At other wavelengths in the scatter dominated region, for Mie-like scattering one might expect the scatter coefficient to vary only slowly with wavelength, and data in skin[14] and other tissues[18] show a gradual decrease with increasing wavelength. The absorption coefficient, measured in brain and muscle tissue in vitro[18] and in brain in vivo[19] shows a broad water peak at around 975 nm, and similar small peaks have been found in skin at around 1400 to 1900 nm[14]. There are no systematic measurements of the anisotropy parameter, g, as a function of wavelength through the scatter dominated region. The effective penetration depths show a slow increase or are essentially constant with increasing wavelengths above 630 nm, with a slight dip corresponding to the 975 nm water peak[20].

Comparable scatter and absorption

In the higher wavelength band around 2000 nm, there are few experimental measurements of the effective penetration in different tissues or of the fundamental scatter and absorption parameters[14]. In the region of UVA and visible light below 600 nm, there are absorption and scattering data for skin, determined indirectly from measurements of diffuse reflectance and transmittance[14,21]. These show a decrease in $\mu_s(1-g)$ from ~30 mm[-1] at 300 nm to ~8 mm[-1] at 600 nm. The absorption coefficient was essentially constant, although it should be noted that this tissue was probably blood-free as measured in vitro. Wilson et al.[18] showed that for brain and muscle, the scatter coefficient decreased only slowly with increasing wavelength from 400 to 630 nm, and that the shape of the scattering spectrum was very similar for the two tissues. The absorption coefficient, by contrast, showed a very marked peak in the green at around 550 nm, and rose sharply again below 500 nm. This structure, which corresponds to the absorption of hemoglobin, also appears in the effective attenuation spectra. The only known data for the wavelength dependence of g are those of Bruls and van der Leun[22] in human epidermis in vitro, for which g increased from ~0.82 to 0.87 as the wavelength increased from 300 to 550 nm. Interestingly, Mie theory predicts that the phase function should, to a first approximation, become less forward peaked (decreasing g) with increasing wavelength.

IMPLICATIONS FOR LIGHT PROPAGATION MODELLING

In this section we shall consider how the optical characteristics found in tissues constrain the propagation models which may be usefully employed to calculate the distribution of light energy fluence in these tissues. Again, it is convenient to consider the three scatter/absorption regimes separately. Note that the discussion here is confined to cases where the optical properties are constant during irradiation and where the local rate of energy absorption is proportional to the product of μ_a and the local energy fluence rate. Non-linear, threshold photophysical processes such as multi-photon absorption are not taken into account.

Absorption dominated

In this case, taking the example of a light beam incident normally on a plane tissue surface, there is little lateral spread of the beam since the scattering is small. The energy fluence rate decreases with depth, x, approximately as a single exponential (Beer's law):

$$\psi(x) = \psi_0.e^{-\mu_a \cdot x} \tag{2}$$

The incident forward peaked radiance is maintained with depth in tissue, and only slowly becomes more isotropic due to scatter. More complex irradiation and tissue geometries are relatively simple to take in account, for example by considering the incident light field as a set of independent pencil beams.

Scatter dominated

This is the case for which most progress has been made in recent years. As shown by the example in Figure 3, which has the same geometry as above, the fluence is no longer a simple exponential function of depth, but may have a significant sub-surface build-up. The radiance pattern reveals the transition from the collimated beam at the surface to a more isotropic distribution at depth. Note, however, that the radiance can still be substantially anisotropic even at several effective penetration depths below the surface.

Much of the initial modelling work for this case considered slab geometries and various approximate methods based on "scoring" the forward and backward light fluxes within layers of the slab. Such models have found particular application in skin optics[14,21,23-25], and have provided a baseline of useful data with which to compare other, more general models. However, this approach becomes more difficult to use as the scattering becomes very forward peaked, or for other irradiation or tissue geometries.

For dosimetry in photodynamic therapy the most widely used model has been diffusion theory (see reviews in refs. 6,7,17,26), which is based on reducing the

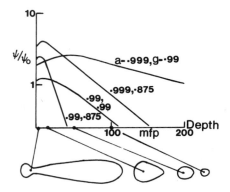

Fig. 3. Energy fluence and (lower diagram) radiance distributions in a highly scattering medium. The radiance plots are for the case of a = 0.999, g = 0.95, and correspond to the depths indicated. These data were calculated using a Monte Carlo photon transport model with Henyey-Greenstein phase functions as in Flock et al.[9,28].

Boltzman equation to a more manageable differential equation in the local energy fluence rate $\psi(\underline{r})$:

$$\nabla^2 \psi(\underline{r}) - \frac{\mu_a}{D} \psi(\underline{r}) = -\frac{S_0(\underline{r})}{D} + 3\underline{\nabla} \cdot S_1(\underline{r})$$

(3)

where the so-called diffusion coefficient, D, is given by $1/D = 3(\mu_a + (1-g)\mu_s)$, and S_0 and S_1 are the first two coefficients in the Legendre expansion of the general source function $S(\underline{r}, \hat{\Omega})$.

The assumptions of diffusion theory in this form are that the "transport albedo" is high ($\mu_a \ll \mu_s(1-g)$) and that the radiance pattern within the tissue is at most linearly anisotropic. Under these conditions, equation 3 can yield simple analytic expressions for many of the quantities of interest. Thus, for example, for a large diameter collimated beam incident on the surface of semi-infinite block of homogeneous tissue, the energy fluence rate at depth is given by

$$\psi \alpha \; e^{-\mu_{eff} \cdot x}$$

(4)

where the effective attenuation coefficient is

$$\mu_{eff} = 1/\text{effective penetration depth}$$
$$= [3 \mu_a (\mu_a + \mu_s(1-g))]^{1/2}$$

(5)

The corresponding radiance distribution at depth is

$$L(\cos \theta) = (1 + 3 \; g(1-a) \; \mu_t \cos \theta / \mu_{eff})/(1 - \mu_{eff} \cos \theta / \mu_t)$$

(6)

Equivalent expressions for the radial dependence of fluence with distance from interstitial spherical or cylindrical fiber sources have been derived (see, e.g. ref. 27).

Such simple analytic solutions obtained from diffusion theory make this an attractive model, but the required assumptions of high albedo and linear anisotropy in the radiance restrict its application. For example, Flock et al.[28] have shown that the calculation of the effective attenuation coefficient becomes decreasingly accurate as the albedo decreases, especially for high g values. The error is around 10% even at a = 0.95, g = 0.9. Further, diffusion theory fails near sources and tissue boundaries since the radiance may be supralinearly anisotropic. There are also limits to the complexity of the source and tissue geometries which can be modelled by diffusion theory, and it is difficult to obtain the absolute normalization of the fluence distribution.

A possible alternative, Monte Carlo simulation, is a computational model in which individual photon trajectories are traced as each photon undergoes successive scattering interactions until it exits the tissue volume or is locally absorbed. The interaction path lengths and the scattering angles are randomly sampled, with appropriate weighting for the scattering and absorption coefficients and angular phase function. This method was first used to study light propagation in tissue by Wilson and Adam[29], who calculated the depth distribution of photon fluence for tissue albedos of 0.2 and 0.8 with isotropic scattering (g=0). Their results are, therefore, more relevant to the third case of comparable absorption and scattering than to the scatter dominated case. Recent extension of this work[28] has examined higher albedo tissues and incorporated anisotropic phase functions of high g value. The method has also been applied to near-infrared spectroscopy[30,31]. Figure 4 illustrates typical fluence distributions obtained by Monte Carlo simulation for pencil beam irradiation in media with scatter dominated optical properties characteristic of those found in tissues at 630 nm.

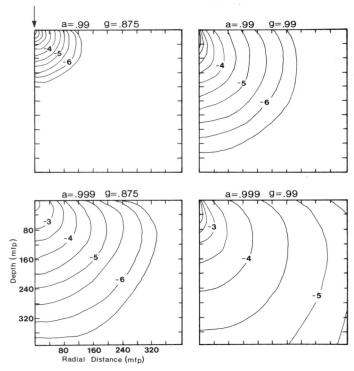

Fig. 4. Iso-fluence distributions for incident pencil beams in highly scattering media for different values of albedo and anisotropy. The vertical axis is the depth in mean-free paths (1 mfp = $1/\mu_t$), and the horizontal axis is the corresponding radial distance from the incident beam entry point. The number on each contour is $\log_{10}(n/N)$ where N = number of incident photons (50,000 in these simulations) and n = number of photons passing through a sphere with cross-sectional area of 1 mfp^2.

NON-INVASIVE MEASUREMENTS OF TISSUE OPTICAL PROPERTIES IN VIVO

The measurements referred to in the section above on the optical properties of tissues have been made either in vitro using tissue samples, or have involved the invasive use of interstitially placed optical fiber detector probes. These studies, which require extension to provide data over a much wider wavelength range, have been valuable in defining the generic characteristics of soft tissues. This knowledge has led to the development and continuing refinement of light propagation models as discussed in the previous section.

However, these methods are not suitable for the determination of tissue optical properties in clinical applications in which one needs to ascertain the values of one or more of the parameters μ_a, μ_s and g (or same equivalent set such as μ_t, a and g or μ_{eff}, a and g) in specific tissues in individual patients. For this, non-invasive or minimally invasive techniques must be used. Further, for most therapeutic applications such as laser surgery or photodynamic therapy, there is usually access only to one surface of the tissue, namely that which is to be irradiated. (This restriction may be less severe in some diagnostic applications, where, for example, IR transmission through a whole organ or body part such as the head or a limb may be feasible (e.g. ref. 35)). Thus, the main measurable quantity will be the diffuse reflectance through the irradiated tissue surface.

As illustrated in Figure 5, diffuse reflectance measurements may be made using 2 main geometries, and we will now speculate on the tissue optical properties which may be obtained from such measurements.

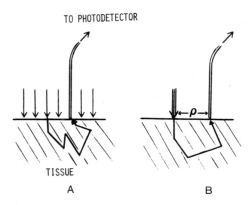

Fig. 5.   Illustration of measurements of A)   the total diffuse reflectance, R, from an irradiated tissue, and B)   the local diffuse reflectance, R(ρ), as a function of radial distance ρ from an incident pencil beam.

The chief advantage of the Monte Carlo method is its complete generality, since there are no a priori constraints on the tissue optical properties or irradiation and tissue geometries. However, for optical properties typical of soft tissues in the therapeutic window, there are considerable practical limits due to the high total attenuation, μ_t, which makes the distance between interactions very small, and the high albedo which requires that many scattering interactions be followed before the photon is absorbed or lost to the volume. Wyman and Patterson[32] and Wyman et al.[33] have examined various ways to reduce the computation time in Monte Carlo modelling in the scatter dominated region, but obtaining reasonable statistics far from sources remains computationally expensive or impracticable.

We have recently proposed[28] that diffusion theory and Monte Carlo modelling may be combined, to yield a hybrid model which retains the strengths and circumvents the limitations of each technique used alone. Thus, for example, Monte Carlo modelling may be used to calculate the fluence distributions close to sources and boundaries where diffusion theory fails, while diffusion theory may be used to extrapolate the spatial distributions at depth within tissue.

A third model, recently proposed by Bonner et al.[30], is based on a random walk formalism whereby the photon distribution is scored on a 3-dimensional lattice, with appropriately weighted probabilities for photon transport from one lattice point to its nearest neighbors. This elegant model produces analytic expressions for many quantities of interest, in the limit of many steps in the random walk. It is thus particularly suited to handle calculations in the scatter dominated regime and is complementary to the Monte Carlo technique which is impracticable for many-scattered photons because of limited statistics.

Comparable scatter and absorption

This third case has been much less studied than the scatter dominated regime. For simple slab geometry some of the established flux-counting methods are applicable. Clearly diffusion theory is largely inappropriate, although more complex methods based on the same idea of simplifying the Boltzman equation to a workable analytic form may be useful, as for example in the work of Star et al.[26]. Monte Carlo techniques will work in this regime, as shown by Wilson and Adam[29]. However, the statistical accuracy will generally be worse than for high albedo for the same total attenuation coefficient. There is also a problem in using this model for the derivation of optical scattering and absorption data from indirect bulk tissue measurements, since there is no analytic "inverse solution", i.e. one cannot work backwards from the measured quantities to the underlying tissue properties. Thus, an iterative procedure[34] is needed to fit the experimental data.

i) The total diffuse reflectance, R, may be measured. However, this does not yield a unique value for any of the fundamental tissue optical properties, as can be seen from the example of Figure 6, where R is plotted as a function of albedo or scattering anisotropy. Any given value of R can result from a range of combinations of a and g.

ii) Alternatively, or in addition to the total diffuse reflectance, the local diffuse reflectance, R (ρ), may be measured as a function of the radial distance, ρ, from the source. This may yield μ_{eff}, but experimental and theoretical studies are needed to investigate whether it will also enable a or g to be separately determined.

iii) It may be possible to add to the tissue a known concentration of an absorbing dye, and to measure R or R(ρ) as a function of this added absorption. (The converse of this has been developed by Patterson et. al.[36] to determine the concentration of photosensitizer in tissue, given the optical properties.) Other in vitro studies of this "added absorber" technique[15]

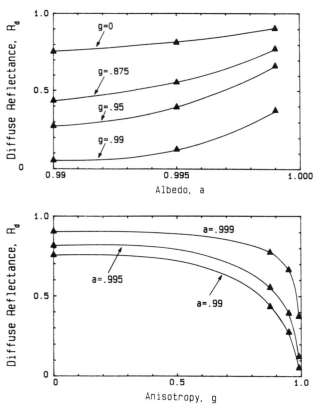

Fig. 6. Total diffuse reflectance, R, calculated by Monte Carlo simulations of 50,000 photons incident normally on a semi-infinite slab of homogeneous tissue with various values of albedo and scattering anisotropy. The lines are drawn by eye through the computed data points.

suggest that μ_a and $\mu_s(1-g)$ might be derived from such measurements. This concept, of course, raises the question of how to determine the dye concentration in the tissue. Non-optical techniques such as nuclear counting with radiolabelled dye have been investigated[37]. Optical methods based on quantitative fluorescence measurements have also been proposed (e.g. ref. 38), but, as with reflectance spectrophotometry, these require a knowledge of some of the optical properties of the tissue in order to achieve absolute quantitation. Thus, the total problem may be underconstrained.

iv)   Finally, with the advent of picosecond lasers and fast time-resolved detection techniques, the additional information from the photon time-of-flight may be of value. For example, Figures 7-9 show Monte Carlo results for R and R(P) as a function of time-of-flight in the tissue, for a range of albedo and scattering anisotropy typical of tissues at 630 nm. Computational limits presently preclude simulations to longer times or larger values of P.

It is seen in Figure 7 that R(t) depends on both a and g, but that in the range applicable to soft tissues in the scatter dominated regime (a > 0.99, g > 0.5), the shape of

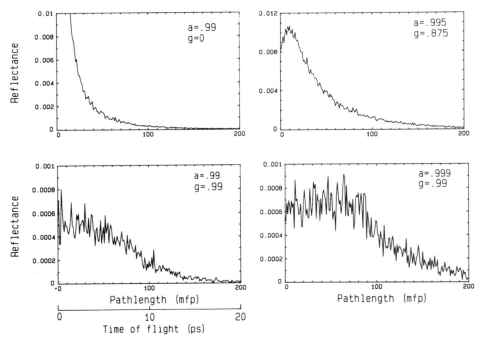

Fig. 7.   Examples from Monte Carlo simulations of the total diffuse reflectance time-of-flight distributions, R(t). The reflectance, R, is plotted as a function of the path length travelled by the photons in tissue before exiting the irradiated tissue surface (pencil beam irradiation geometry). The equivalent time-of-flight is based on a typical mean free path of around 30 μm for soft tissue at 630 nm[9], with no correction for the tissue refractive index.

the time spectrum changes more with anisotropy than with albedo. As illustrated in Figure 8, at very short times R(ρ, t) also shows a strong dependence on g but much lesser dependence on albedo. Conversely, at long times, say > 20 ps, the total counts in R(ρ, t) for ρ > 10 mfp increase with increasing albedo for the same anisotropy. Further, it can be demonstrated using the model of Bonner et al.[30] that, under certain geometric and optical conditions, the dependence of R or R(ρ) on time, at long times following an incident light pulse, may yield the tissue absorption coefficient, μ_a, independent of the scattering (Patterson, et al., in preparation). These preliminary observations suggest the hypothesis that the time spectrum of reflectance could be used to determine separately μ_a, μ_s and g (or some equivalent set). This will be tested in future modelling and experimental studies.

CONCLUSIONS

We have shown that the fundamental optical scattering and absorption properties of tissues not only profoundly affect the spatial distribution of light in the tissues, as expected, but also strongly influence the development of theoretical quantitative models of light propagation and the techniques which will be required to measure these optical properties in vivo. There is indeed much positive synergism between the current research on theoretical modelling and development of experimental techniques. This can be considered a healthy sign for future progress in light dosimetry in tissue.

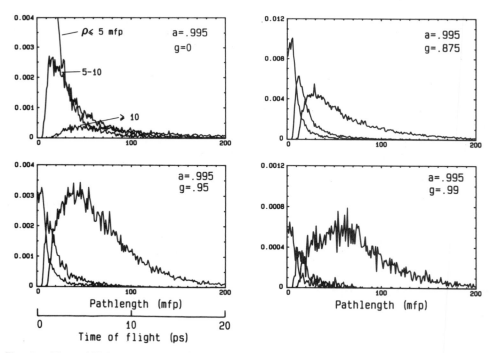

Fig. 8.   Time-of-flight spectra for R(ρ) for 3 ranges of ρ, fixed albedo of 0.995 and 4 different values of anisotropy.

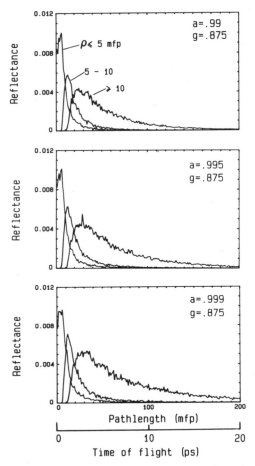

Fig. 9. As for Figure 8, with fixed anisotropy and varying albedo.

ACKNOWLEDGEMENT

The experimental and theoretical work on light dosimetry at this institution is supported by the National Cancer Institute of Canada.

REFERENCES

1. D.R. Doiron, L.O. Svaasand and A.E. Profio, Light dosimetry in tissue: application to photoradiation therapy, in: "Porphyrin Photosensitization," D. Kessel and T.J. Dougherty, eds., Plenum, New York (1983).
2. J.P.A. Marijnissen and W.M. Star, Phantom measurements for light dosimetry using isotropic and small aperture detectors, in: "Porphyrin Localization and Treatment of Tumors," D.R. Doiron and C.J. Gomer, eds., Liss, New York (1984).

3. L.O. Svaasand, Optical dosimetry for direct and intersitial photoradiation therapy of malignant tumors, in: "Porphyrin Localization and Treatment of Tumors," D.R. Doiron and C.J. Gomer, eds., Liss, New York (1984).

4. B.C. Wilson, W.P. Jeeves, and D.M. Lowe, In vivo and post-mortem measurements of the attenuation spectra of light in mammalian tissues, *Photochem. Photobiol.,* 42:153-162 (1985).

5. J.P.A. Marijnissen, W.M. Star, J.L. van Delft and N.A.P. Franken, Light intensity measurements in optical phantoms and in vivo during HPD-photoradiation treatment, using a miniature light detector with isotropic response, in: "Photodynamic Therapy of Tumors and Other Diseases," G. Jori and C. Perria, eds., Edizioni Libreria, Padova (1985).

6. W.M. Star, J.P.A. Marijnissen, H. Jansen, and M.J.C. van Gemert, Light dosimetry: status and prospects, *J. Photochem. Photobiol.,* B1:149-167 (1987).

7. M.S. Patterson, B.C. Wilson and D.R. Wyman, The propagation of optical radiation in tissues, in: "Advances in Laser Biophysics," M.J. Colles, ed., JAI Press, England (in press) (1989).

8. A. Ishimaru, "Wave Propagation and Scattering in Random Media," Academic, New York (1978)

9. S.T. Flock, B.C. Wilson and M.S. Patterson, Total attenuation coefficients and scattering phase functions of tissues and phantom materials at 633 nm, *Med. Phys.,* 14:835-841 (1987).

10. J.J. Duderstadt and L.J. Hamilton, "Nuclear Reactor Analysis," Wiley, New York (1976).

11. J.L. Boulnois, Photophysical processes in recent medical laser developments: A review, *Lasers Med. Sci.,* 1:47-66 (1986).

12. B.C. Wilson, M.S. Patterson and S.T. Flock, Indirect versus direct techniques for the measurement of the optical properties of tissues, *Photochem. Photobiol.,* 46:601-608 (1987)

13. S.L. Jacques, C.A. Alter and S.A. Prahl, Angular dependence of HeNe laser light scattering by human dermis, *Lasers Life Sci.,* 1:309-313 (1987).

14. R.R. Anderson, J. Hu and J.A. Parrish, Optical radiation transfer in the human skin and applications in in vivo remittance spectroscopy, in: "Bioengineering and the Skin," R. Marks and P.A. Payne, eds., MTP Press, Lancaster (1981).

15. B.C. Wilson, M.S. Patterson and D.M. Burns, Effect of photosensitizer concentration in tissue on the penetration depth of photoactivating light, *Lasers Med. Sci.,* 1:235-244 (1986).

16. J.E. Tyler, Optical properties of water, in: "Handbook of Optics," W.G. Driscoll and W. Vaughan, eds., McGraw-Hill, New York (1978).

17. B.C. Wilson and M.S. Patterson, The physics of photodynamic therapy, *Phys. Med. Biol.,* 31:327-360 (1986).

18. B.C. Wilson, M.S. Patterson, S.T. Flock and J.D. Moulton, The optical absorption and scattering properties of tissues in the visible and near-infrared wavelength range,

in: "Light in Biology and Medicine," R.H. Douglas, J. Moan and F. dall'Acqua, eds., Plenum Press, New York (1988).

19. S. Wray, M. Cope, D.T. Delpy, J.S. Wyatt and E.O.R. Reynolds, Characterization of the near infrared absorption spectra of cytochrome aa_3 and hemoglobin for the noninvasive moitoring of cerbral oxygenation, *Biochm. Biophys. Acta*, 933:184-192 (1988).

20. F.P. Bolin, L.E. Preuss and B.W. Cain, A comparison of spectral transmittance for several mammalian tissues: effects at PRT frequencies, in: "Porphyrin Localization and Treatment of Tumors," D.R. Doiron and C.J. Gomer, eds., Liss, New York (1984).

21. W.J.M. van der Putten and M.J.C. van Gemert, A modelling approach to the detection of subcutaneous tumors by hematoporphyrin-derivative fluorescence, *Phys. Med. Biol.*, 28:639-645 (1983).

22. W.A.G. Bruls and J.C. van der Leun, Forward scattering properties of human epidermal layers, *Photochem. Photobiol.*, 40:231-242 (1984).

23. M.J.C. van Gemert and W.M. Star, Relations between the Kubelka-Munk and transport equation models for anisotropic scattering, *Lasers Life Sci.*, 1:287-298 (1987).

24. G. Yoon, A.J. Welch, M. Motamed and M.J.C. van Gemert, Development and application of three-dimensional light distribution model for laser irradiated tissue, *IEEE J. Quant. Electr.*, QE-23:1721-1733 (1987).

25. M.J.C. van Gemert, G.A.C.M. Schets, M.S. Bishop, W.F. Cheong and A.J. Welch, Optics of tissue in a multi-layer slab geometry, *Lasers Life Sci.*, 2:1-18 (1988).

26. W.M. Star, J.P.A. Marijnissen and A.J.C. van Gemert, Light dosimetry in optical phantoms and in tissues: I. Multiple flux and transport theory, *Phys. Med. Biol.*, 3:437-454 (1988).

27. A.L. McKenzie, How many external and interstitial illumination be compared in laser photomdynamic theory?, *Phys. Med. Biol.*, 30:455-460 (1985).

28. S.T. Flock, B.C. Wilson and M.S. Patterson, Hybrid Monte Carlo-diffusion theory modelling of light distributions in tissue, *SPIE Proc.*, 408 (in press) (1988).

29. B.C. Wilson and G. Adam, A Monte Carlo model for the absorption and flux distributions of light in tissue, *Med. Phys.*, 10:824-830 (1983)

30. B.F. Bonner, R. Nossal, S. Havlin and G.H. Weiss, Model for photon migration in turbid biological media, *J. Opt. Soc. Am.*, A4:423-432 (1987).

31. P. van de Zee and D.T. Delpy, Simulation of the point spread function for light in tissue by a Monte Carlo technique, *Adv. Exp. Med. Biol.*, 215:179-192 (1987).

32. D.R. Wyman and M.S. Patterson, A discrete method for anisotropic angular sampling in Monte Carlo simulations, *J. Comp. Phys.*, (in press) (1988).

33. D.R. Wyman, M.S. Patterson and B.C. Wilson, Similarity relations for anisotropic scattering in Monte Carlo simulations of deeply penetrating neutral particles, *J. Comp. Phys.*, (in press) (1988).

34. P.A. Wilksch, F. Jacka and A.J. Blake, Studies of light propagation through tissue, in: "Porphyrin Localization and Treatment of Tumors," D.R. Doiron and C.J. Gomer, eds., Liss, New York (1984).

35. D.T. Delpy, M.C. Cope, E.B. Cady, J.S. Wyatt, P.A. Hamilton, S. Hope and E.O.R. Reynolds, Cerebral monitoring in newborn infants by magnetic resonance and near infrared spectroscopy, *Scand. J. Clin. Lab. Invest.*, 47, Suppl. 188:9-17 (1987).

36. M.S. Patterson, B.C. Wilson, J.W. Feather, D.M. Burns and W. Pushka, The measurement of dihematoporphoyrin ether concentration in tissue by reflectance spectrophotometry, *Photochem. Photobiol.*, 46:337-343 (1987).

37. B.C. Wilson, G. Firnau, W.P. Jeeves, K.L. Brown and D.M. Burns-McCormick, Chromatographic analysis and tissue distribution of radiocopper-labelled hematoporphyrin derivatives, *Lasers Med. Sci.*, (in press) (1988).

38. A.E. Profio and D.R. Doiron, Transport of light in tissues in photodynamic therapy, *Photochem. Photobiol.*, 46:591-599 (1987).

SOME PROSPECTS FOR ADAPTING FLUORESCENCE INSTRUMENTATION

Jay R. Knutson

Laboratory of Technical Development
National Heart, Lung and Blood Institute
Building 10, Room 5D-10
Bethesda, MD 20892

We can classify the information content of light which is emitted, reflected or scattered by tissue as four types: 1, direction of propagation; 2, polarization state; 3, energy (wavelength) and 4, time of detection. Unfortunately, the first two are rather quickly "scrambled" in turbid media. Scattering also broadens energy lines and disperses path length; hence all of these may be degraded.

The prospects for probing dense tissues with light are thus quite uncertain (and a motivation for this workshop). One certainty emerges: Association of several measured features will provide more information than any single parameter. In the realm of fluorescence spectroscopy, it is well established that mathematical associations between two or more of data types 2, 3 & 4 yield useful signatures and improve kinetic rate recoveries [1,2 and references therein]. Propagation direction is usually fixed by instrumental geometry and linked by symmetry to polarization.

In conventional scattering spectroscopies, some restrictions on geometry and (heterodyne) energy selection are found useful, but direct temporal resolution has been reserved for pulse-probe events or background suppression for Raman spectroscopy[3,4].

If temporal resolution is going to emerge in diffuse scattering studies, it seems likely that some of the technology can be "transplanted." The adaptation of Holtom's TCSPC (time-correlated single-photon counting) system presented here reinforces that hope.

In this brief note, I will outline some multichannel detection schemes that have been constructed (or are being prototyped now) at NHLBI-LTD. These schemes were previously presented along with some intended for protein folding[5] and those proceedings are recommended as an overview of time resolved fluorescence that include more diverse views and topics.

Our focus has centered on DAS (decay associated spectra) that provide spectral signatures for each decay function (often a simple exponential lifetime) that is present in a mixed emission surface[6]. This priority led to the construction of the detector assembly in Figure 1. The manufacturer had developed the multi-anode microchannel plate photomultiplier tube (MAMCP-PMT) for imaging applications neglecting time resolution; our inquiries led them to test the standard tube for transit time spread (TTS); their ca. 100ps results gave us confidence to proceed. Meanwhile, they modified anode terminations and delivered a tube capable of <60 ps TTS, quite comparable to the single anode tubes of similar (12-micron-channel) multiplying wafer construction. They have achieved this level of performance repeatedly[7].

Unfortunately, the present configuration yields TTS of 250ps. About half of that can be assigned to simple dispersion in our polychromator (along with other optical delays; see Holtom[8]). The remainder we ascribe to the use of outmoded discriminators and amplifiers now being replaced with faster units. After the update of electronics is complete, we expect ca. 100ps TTS which can (via iterative reconvolution) provide reliable data on decays as short as 15-20ps. For more stringent needs, we will need to replace the monochromator with subtractive designs[8] or wedged interference filters.

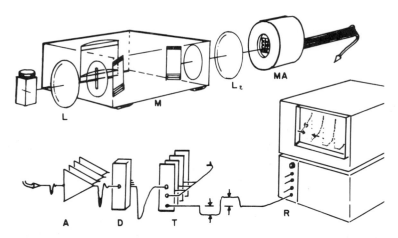

Fig.1(a). Multiple Anode MCP-PMT (MA) for simultaneous collection of decay data at several wavelengths. Luminescence exiting sample after pulsed excitation is collected by lens (L) and dispersed in momochromator (M) configured as polychromator (exit slit absent). Relay lens (L2) images slit plane across anodes. Photoelectron-induced pulses from each anode are amplified (A), constant-fraction discriminated (D) and timed with a TAC (time-to-amplitude converter). Pulse heights are analyzed in individual memory segments of a routed multichannel analyzer (R).

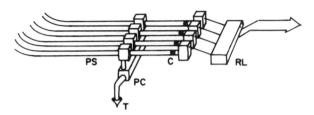

Fig. 1(b). Alternative to 1a. PMT pulses are "grouped" with an array of power splitters (PS) whose first output is multiplexed via power combiner (PC) into a single timing channel (T). The second output of each splitter is used to fire a simple comparator to control routing logic.

In our system, simultaneous measurement at several wavelengths was desired; the standard 4 x 4 array is less convenient for this, although Bowman [pers. comm.] noted the ease with which slit-plane position could be passed through fiberoptic ribbons. Ross and Laws [pers. comm.] noted, in contrast, that Wollaston prism coupling would expand the value of the tube to another parameter at the same time. In any case, the system cost per channel is a scaling problem, since each requires an amplifier, discriminator and TAC. Fortunately, all but the latter are sold in packages of 4, 8, or 16 for NIM multiplexing needs. To keep this in perspective, each channel adds only about 2% to the cost of commercial synch-pumped, cavity-dumped dye laser systems.

The information available from photons diffusing out of tissue masses is of unknown and controversial value at present, but the utility of multiple detection points is likely to be high even if tomographic detail cannot be expected. If for no other reason, the multiple measurements can monitor for exclusion any special sensitivity to proximate boundaries (to be discussed later).

The improbability of complete tissue penetration has been clearly established, but it is revealing to consider the power of direct temporal resolution by photon counting. Assume for the moment we only need a 10% accurate absorbance; a few hundred photons will suffice. Each pulse of the dye laser yields over a billion photons (even if the 30nj pulses are poorly coupled). Better yet, these pulses strike the tissue at megahertz rates. In a few seconds, then, we accumulate a discrimination ratio for prompt vs. delayed of 10^{14} !! By this we mean that some 10^{16} photons are incident in ten seconds, and we need measure only a hundred or so arriving very early. Their early arrival makes moot the issue of TAC "pulse pileup"[6], until the multiplier saturates. Even that problem can be reduced with judicious gain gating. Better yet, photon counting is a relatively "background-free" measurement, so accuracy will continue to grow like (time$^{1/2}$) for many decades. A cooled PMT may require hundreds of seconds to register a single random dark count in the narrow (subnanosecond) time window proposed.

In this workshop, strong demonstration of gating efficacy in less turbid solutions was shown, with clear images. Given the sensitivity of TCSPC, it seems likely that many, many more layers of scattering can be pierced.

The sufficiency of TCSPC in real tissues is testable using single channel detection at present; just as confocal scanning microscopy uses only one detector, so might this image problem (eg. via galvonometric or AO scanning). The MAMCP-PMT shown above is also tempting, although a more direct solution might emerge from another detector we have proposed (see Figure 2). In this system, a single timing circuit can be used to select temporal "slices" for memory classification, while image position of each photon is registered by conventional PSD circuit. A counter-scheme originally proposed by Holtom, using capacitive coupling in a commercially available tube (lacking the timing grids we proposed) to extract "stop" pulses, has already found success to ca. 300ps[9]. The timing grids should return us to intrinsic limits near 50ps TTS.

In time-resolved fluorometry, both pulse TCSPC and phase/modulation methods have been popular, with the latter improving greatly in this decade [see Lakowicz, this volume,

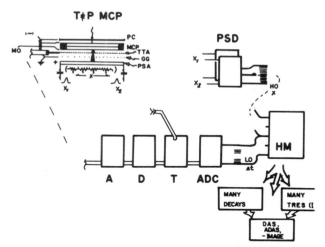

Fig. 2. Time & position MCP-PMT concept. A portion of the photoelectron burst exiting the microchannel plate multiplier (MCP) is intercepted by a timing grid (TTA). The remainder traverses guard grid (GG) enroute to position-sensing anode (PSA). The charge distribution reaching x1 and x2 can be used to compute x, the charge "center-of-gravity," hence selecting high-order bits. The low-order bits are selected by conventional (compare Fig 1a.) timing electronics with an analog-to-digital converter (ADC) or SCA, and buffered into histogramming memory(HM). Hence photon arrival time and arrival position both control address lines for the accumulation of multidimensional intensity (count) profiles.

for review]. At present, the timing stability of heterodyne detection has made phase fluorometry a leader in extracting decays much faster than the TTS. Basically, the two methods have similar sensitivity, although the analog detection inherent in mixing sets up a requirement for about an order of magnitude more intensity. In addition, the sort of information obtained in the frequency domain must be transformed to yield time "windows" that are directly provided by a TAC (time-amplitude conversion). In all, both methods have merit , since the high throughput of phase data makes general time-resolved images easy to compute [E. Gratton, pers.comm.]. For this particular application, it will depend on whether we wish to focus on "early gated" photons or the decay profile of the entire diffusive process.

The results presented at this workshop by Chance & coworkers provide compelling evidence for a decay process that reports the absorbance changes in the diffusive body as changes in a quasi-exponential decay rate (in accordance with what one might expect from Beer's law in a uniform environment). They also showed some variation of the rate with position on the body. In this regard, it is interesting to forgo complex simulation and examine simple analytical formulae. In this way, we can develop some physical intuition for the influence of boundary conditions. To begin we can look at the Green's function for conduction about a point impulse source in a non-absorbing medium:

$$I(r,t) \;=\; (1/4\pi K t)^{3/2} \quad \exp(-r^2/4Kt)$$

where r is distance, t time and K is conductivity (or D).

While this function is clearly not a simple exponential, a log plot reveals conditions where it appears to be exponential. A simple BASIC simulation is shown in Figure 3. In these plots, the log of photon concentration (eg., log intensity) for a family of curves is plotted against a=4Kt, where K is the conduction/diffusion constant. As explained in [10], K may be calculated as b'c/3, where b' is the effective scattering length (on the order of 5X the isotropic mean free path) and c is the speed of light. The different curves represent different values of r^2, the square of the distance from source to observation. As expected, curves from larger r^2 values peak later and at lower intensity values. The units of the plots were intentionally unspecified to permit scaling. For example, a typical plot extends to r^2 range of 1000, so r_{max} is about 31. This makes mm a convenient unit for arm or leg studies. To translate the x axis, one may choose b'. For example, if b'=200u (imfp=40u), then one unit of a is 12.5 ps. The dot spacing (10a) then corresponds to 125ps, and a plot spanning 2000a is 25ns wide. The unitless scaling is adaptable; ie., b' of 20u implies a width of 250ns. The simple QuickBasic program is presented in Appendix.

More important is the influence of proximate absorbing boundaries (remember that light exiting the diffusive body is very unlikely to return, hence the absorbing condition makes sense). The Green's function for a point source near an absorbing plane is:

47

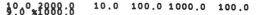

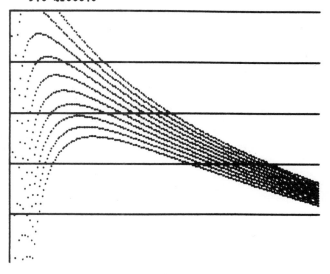

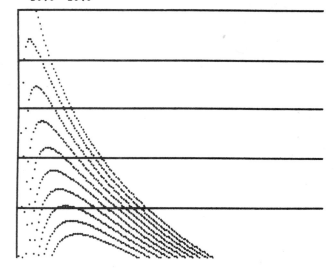

$$I(x,y,z,t) = (1/4\pi Kt)^{3/2} \quad *[$$
$$\exp(-((x-f)^2 + y^2 + z^2)/4Kt)$$
$$-\exp(-((x+f)^2 + y^2 + z^2)/4Kt) \quad]$$

where a is the distance along x to the plane absorber. Note that this solution uses the famous "image source method" to obtain the black boundary (cf.[11] Mathews & Walker, p. 245).

The BASIC simulation for this case is interesting in that the quasi-exponential region is more apparent. More important, the slope in the log plot (the apparent decay rate) is quite sensitive to (f), the proximity of the boundary when compared to the other dimensions.

10.0 2000.0   10.0 100.0 1000.0 100.0
10.0 40.0

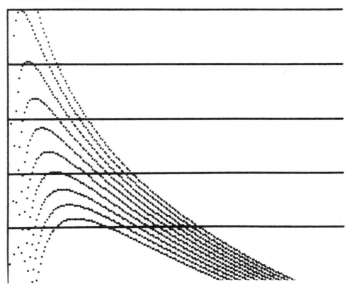

Fig. 3. Simulated conduction-controlled decay profiles. The heading numbers are x axis units of "a" as a_{min}, a_{max}, $a_{increment}$, followed by a set of r2 values (min,max,inc). The second line lists "f" offset in r2 units (see text) to absorber along with attenuation (log units). These semilog plots mimic emission decays, with "a" = 4Kt scaling the horizontal axis. A typical curve is 25ns wide (see text). The different plots demonstrate importance of "f" (eg., the proximity of open space = absorbing boundary) to rate of intensity decay, since other parameters are the same.

It is difficult to know how important an absorbing "sink" will be in clinical applications, but it seems encouraging that the decay profile can be drastically altered. We have found that the time-resolved luminescence of complex systems is best reconciled by "global analysis," ie. the simultaneous analysis of many decay profiles for a set of parameters common to them[12]. Even though the diffusive system is more complex, it seems likely that a series of curves taken at different wavelengths, at different spatial positions, with different pulse structures, etc., will combine well in the computer to yield more than average absorbance.

The complexity of this data-analysis problem is such that image charge and other analytical formulae will quicky fail. One promising approach is recursive simulation for nonlinear least-squares estimation. Rather than attempt the full mesh simulations shown in the workshop, one might begin by coupling a series of "boxes" that comprise the body in question. For example, a group of cm-edged cubes can be started with to describe a tissue. The transport between compartments might be presumed isotropic (and each compartment would be treated as uniform by definition). The eigenvector-eigenvalue decomposition of coupled compartments is well known from pharmacokinetic and other fields (see [13] and references therein). If a series of different exit points is monitored simultaneously and compared with simulation, even the nanosecond data may prove valuable. As parameter estimates (the absorbance, etc. of each "box") improve, the size might be refined. Of course, the identifiability of the mesh is not good if we are permitted a wide range of starting estimates, and only experience could tell if some boxes yield identical influence on the observed decay. Given the wide range of geometry available, it seems likely that some means could be found to avoid accidental degeneracies. Similar finite mesh simulations, that as little as ten years ago required mainframe power, now are routinely accomplished on the desktop.

ACKNOWLEDGEMENTS

Thanks to R.L.Bowman for encouraging instrument development as a priority, and for helpful discussions; to G.Holtom, R.F.Chen, Wm.Laws, J.B.A.Ross, L.Brand and members of his lab, for helpful advice and discussions, and to the organizers (tausend tak) for opening an interesting topic and for their kind invitation.

REFERENCES

1. L. Brand, J.R. Knutson, L. Davenport, J.M. Beechem, R.E. Dale, D.G. Walbridge, and A.A. Kowalczyk, Time resolved fluorescence spectroscopy: Some applications of associative behaviour to studies of proteins and membranes, in: "Spectroscopy and the Dynamics of Molecular Biological Systems," P.M. Bayley and R.E. Dale, eds., Academic, London (1985).

2. J. R. Knutson, L. Davenport, J.M. Beechem, D.G. Walbridge, M. Ameloot, and L. Brand, Associated spectra and the multidimensional nature of fluorescence spectroscopy, in: "Excited-State Probes in Biology and Medicine," A. Szabo and L. Masotti, eds., Plenum, New York (in press).

3. R.P. Van Duyne, D.L. Jeanmaire and D.F. Shriver, Mode locked laser raman spectroscopy - A new technique for the rejection of interfering background luminescence signals, Anal.Chem., 46:213 (1974).

4. J.M. Harris, R.W. Chrisman, F.E. Lytle and R.S. Tobias, Sub-nanosecond time-resolved rejection of fluorescence from raman spectra, Anal. Chem., 48:1937 (1976).

5. J.R. Knutson, Time-resolved laser spectroscopy in biochemistry, fluorescence detection: Schemes to combine speed, sensitivity and spatial resolution, J. Lakowicz, ed., SPIE Vol. 909, Bellingham (1988).

6. J.R. Knutson, D.G. Walbridge and L. Brand, Decay-associated fluorescence spectra and the heterogeneous emission of alcohol dehydrogenase, Biochemistry, 21:4671 (1982).

7.   Hamamatsu literature "Applications of Multichannel Plates" and pers.comm. from D. Fatlowitz.

8.   G. Holtom, this volume and pers.comm.

9.   W.G. McMullan, S. Charbonneau and M.L.W. Thewalt, Simultaneous subnanosecond timing information and 20 spatial information from imaging photomultiplier tubes, *Rev.Sci.Inst.* 58(9):1626-1628 (1987).

10. H. Leelavathi, J.P. Pichamuthu, Propagation of optical pulses through dense scattering media, *Appl.Opt.*, 27(12):2461-2468 (1988).

11. J. Mathews and R.L. Walker, "Mathematical Methods of Physics," W.A.Benjamin, New York (1970).

12. J.R. Knutson, J.M. Beechem and L. Brand, Simultaneous analysis of multiple fluorescence decay curves: A global approach, *Chem.Phys.Lttr.*, 102(6):501- 507 (1983).

13. J.M. Beechem, M. Ameloot, J.R. Knutson and L. Brand, The global analysis of fluorescence intensity and anisotropy decay data: Second generation theory and programs, in: "Fluorescence Spectroscopy, Vol.I, Principles and Techniques," J.Lakowicz, ed., Plenum, New York (in press).

APPENDIX

SCREEN 1

```
INPUT amin, amax, ainc, r2mn, r2mx, r2i, slog, xof
LPRINT USING " ####.#"; amin, amax, ainc, r2mn, r2mx, r2i
LPRINT USING " ###.#"; slog, xof
ff$ = CHR$ (12)
LPRINT ff$
CLS
FOR r2 = r2mn TO r2mx STEP r2i
amx = amax - amin
100 FOR a = amin TO amax STEP ainc
q = EXP (-r2 / z) - EXP (-(r2 + xof) / a)
r = (1 / a) ^^1.5
b = q * r
IF (b <= 0) THEN GOTO 111
z = LOG (b)
111 yr = ((z + slog)
```

Part 2.   Experimental Studies

THE NEONATE BRAIN (NIR) AND BREAST IMAGING USING TRANSILLUMINATION

Y. Yamashita, S. Suzuki, S. Miyaki, and T. Hayakawa

Hamamatsu Photonics K.K.
1126-1 Ichino-cho, Hamamatsu City, Japan 435

INTRODUCTION

In physiological and biochemical observations, the ideal probes should allow effective yet non-invasive interaction with the subject, while conveying information to the observer without being affected by surrounding matter. Near infrared (NIR) spectroscopy realizes this dream, by allowing us to make repeated or continuous diagnosis of the human body without harm to the subject. This is achieved by observing spectral responses in tissues which vary according to their biological states. This technique has been used for monitoring the brain in neonates, and also for the diagnosis of breast cancer.

The system for monitoring cerebral oxygenation is capable of detecting absorbance changes of 0.02 O.D. with a background attenuation of 10 O.D. at intervals of 2 seconds. When used in the breast cancer imaging system, images of the whole breast of a remarkable clarity were obtained by scanning a narrow-beam light source and a collimated detector over the breast simultaneously.

This paper is a report on the design and performance of these systems, along with some clinical data.

CEREBRAL OXYGENATION MONITORING SYSTEM FOR NEWBORN INFANTS

Jöbsis first reported that human tissues are relatively transparent to NIR light, and that the intensity of light thus transmitted is dependent on the oxygenation state of the tissue[1]. This is due to the presence of two natural chromophores in this wavelength region which absorb NIR light depending on the amount of oxygen absorbed. These two compounds are hemoglobin (an indicator of blood oxygenation) and cytochrome aa_3 (an indicator of tissue oxygenation).

Several investigators[2,3] have shown that the cerebral oxygenation of infants can be measured using NIR spectroscopic techniques.

System design[4]

Fundamental studies[5,6] on transillumination spectrophotometry of the brain were carried out by D.T. Delpy and his colleagues at University College London. An algorithm was constructed allowing calculation of quantitative changes in oxygenated and deoxygenated hemoglobin and cytochrome aa_3 in the brain.

Given both the order of magnitude of the absorption changes, and the resolution necessary for clinical purposes, the system has been designed and constructed to be capable of measuring absorption changes of less than 0.02 O.D. against a background absorption of 10 O.D.. Figure1 is a block diagram of the system, the important elements of which are discussed below.

In order to be able to measure absorption changes across a tissue as thick as possible, high-power pulsed laser diodes are employed as the light source, and a photomultiplier tube is operated in the photon counting mode. The laser diodes (Laser Diode Labs LA68) can produce a peak optical power of approximately 10W with a pulse duration of 100ns. In order to increase the accuracy of the analysis, six laser diodes ranging from 775nm to 905nm are employed. Because the output of the laser diodes varies with temperature, a small amount of the output of each laser diode is monitored by a photodiode and this data is used in correcting the photon count at each wavelength.

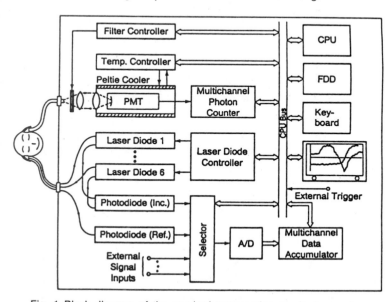

Fig. 1 Block diagram of the cerebral oxygenation monitoring system.

Light reflected by the subject (the head) is also monitored by a second photodiode through additional fibers contained in a transmitting fiber bundle in order to monitor the coupling between the transmitting fiber bundle and the skin. If the signal indicates that the fiber has become detached, the laser diode drivers are automatically disabled.

The photomultiplier tube (PMT) used in the system is HAMAMATSU R943-02 with a gallium arsenide photocathode, giving a high quantum efficiency in the NIR region. To reduce the dark noise, the PMT is mounted in a Peltier-cooled housing and is operated at a temperature of around 5°C (deviation of less than ± 0.1°C).

In addition, a long wavelength pass filter (3dB at 750nm) has been incorporated into the housing to reduce ambient visible light. After shaping and amplification, the PMT signal is sent to a multichannel photon counter.

The timing diagram of the laser diodes and the photon counting operation is shown in Figure 2. Six laser diodes, each pulsed at 4KHz, are repeatedly fired in sequence. At each firing, the multichannel photon counter is operated for a collection time of 250ns. This count rate is optimized by varying the output of each laser diode, or by inserting neutral density filters in the PMT housing. The collected photons are stored in six different channels, with a seventh being used to measure the background count when no lasers are triggered. This background count, subtracted from the total signal and background count, is collected on each of the first six channels.

The system can accept an external trigger signal, and synchronous measurement is available for further applications (as shown in Figure 3). The maximum external trigger rate is 10Hz, and 32 data sets may be obtained in each interval.

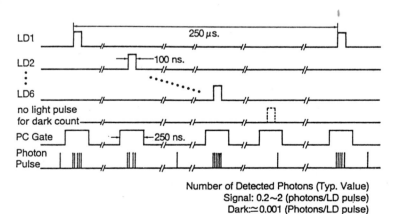

Number of Detected Photons (Typ. Value)
Signal: 0.2~2 (photons/LD pulse)
Dark:≃0.001 (Photons/LD pulse)

Fig. 2. Timing diagram of laser diodes and photon counter in the system.

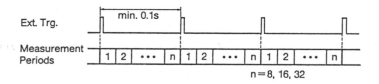

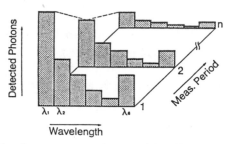

Fig. 3. Synchronous measurement mode with external trigger in the system.

System performance

Initial tests have been carried out to assess the system's linearity, resolution and stability.

Photon counting linearity of the system was measured by adding neutral density filters of small values on a fixed attenuation of 10 O.D.. The results are shown in Figure 4. In this figure, the horizontal axis shows the attenuation changes, and vertical axis shows the number of photons detected per laser pulse. By decreasing the value of neutral density filters, good linearity is achieved up to about 2 photons/pulse, while dark count rate is about 0.001 of a count in the same period.

Resolution of the system is better than 0.02 O.D. against a background attenuation of 10 O.D., with a drift rate of less than 0.004 O.D./hour. Statistically, to obtain a standard deviation of 0.01 O.D., 10,000 photons must be collected in the measuring period, and this takes about 2 seconds on the system if the count rate is set to 1 per pulse. A lower statistical deviation, and hence greater accuracy, can be achieved by collecting for a longer period.

Results

Preliminary tests with the system were performed measuring absorption changes in a resting adult forearm. A cuff around the forearm was used to alter the blood supply. Figure 5 shows the calculated oxygenated hemoglobin and deoxygenated hemoglobin changes. Spectral curve fitting using the least square error technique is used to fit the three unknown compounds (oxygenated hemoglobin, deoxygenated hemoglobin and oxygenated minus deoxygenated cytochrome aa_3) to the six wavelengths. After five minutes the cuff is rapidly

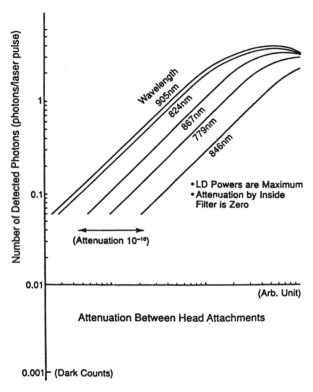

Fig. 4 Photon counting characteristics of the system.

inflated to a pressure sufficient to occlude arterial blood flow. In this situation, the total blood volume remains reasonably constant while the hemoglobin changes from the oxygenated to the deoxygenated state. On release of the cuff, the blood volume increases dramatically, most of the increase being arterial blood, and the amount of oxygenated hemoglobin rises transiently.

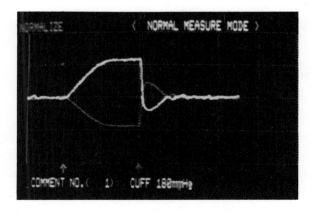

Fig. 5 Changes in the concentrations of HbO_2 and HbR in a resting adult forearm.

This system is currently undergoing trials in the Neonatal Unit at University College Hospital, London. In the normal babies in whom changes in SaO_2 within the range 70 to 97% at constant $PaCO_2$ were observed, a rise in arterial saturation caused a consistent rise in oxygenated hemoglobin and a fall in deoxygenated hemoglobin. Figure 6 shows the marked changes in NIR indices on this study.

Studies in the Neonatal Unit at the University College Hospital, London have shown the feasibility of this system to provide continuous monitoring of changes in cerebral oxygenation.

TRANSILLUMINATION IMAGING ON THE BREAST

X-ray mammography is the most reliable and presently the preferred method for detection of breast cancers[7]. However, due to the carcinogenic effects of exposure to ionizing radiation, alternative modalities are required. Recently, the use of light as a nonionizing probe has been attracting attenuation[8,9,10]. The principle of transillumination is that different tissues display different absorption spectra depending on their biological states[9]. Therefore light can be used to define the localization and to identify a mass according its absorption spectrum. Any optical imaging method, however, encounters difficulties due to high light scattering within the tissues[11,12,13]. In order to overcome this problem, the detecting system must be able to reduce scattered light and multiply the weak transmitted light.

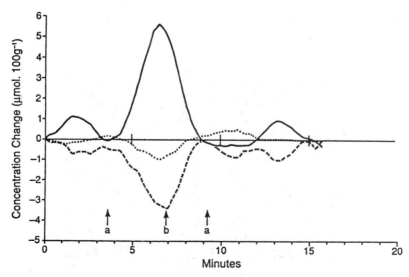

Fig. 6 Changes in the concentrations of HbO_2 (---), HbR (——), and oxygenated cytochrome aa_3 (···) in a preterm infant.

Light scattering within the breast

Light passing through the breast is modified in several ways, including scattering and absorption. These two effects are of primary importance when producing the projected image[14]. Figure 7 (left) demonstrates some possible pathways of light within the breast. A projected image produced by an object far from the surface is strongly affected by the scattered light, resulting in an image with a low contrast between the lesion and surrounding tissue (Figure 7 (right)), making diagnosis more difficult.

Basic principles of the detector system used in the breast cancer imaging system

As mentioned above, light scattering contributes significantly to degradation of the images of human tissues or organs, and it is therefore important that it be reduced. To achieve this, our system employs two fiber collimators; one of which emits collimated laser light, and the other leads mainly the unscattered light to the detector. These two fiber collimators are synchronously scanned over the object to obtain the transillumination images.

Fundamental studies were carried out to estimate the effects of the collimated light source and detector system[15] as shown in Figure 8. Fresh milk was used as the scattering medium, so that scattering would be much greater than in human tissue. The overall path length was 4cm, and the diameter of the wire to be imaged was 2.6mm. The container was scanned in the direction of the Y-axis using a translation stage to obtain the transilluminated image of the wire.

The top row of pictures in Figure 9 shows the images obtained with a broad white light source and an SIT TV camera. The intensity profiles of the images are also traced in these pictures. The wire was imaged at decreasing distances from the detector. Clearly,

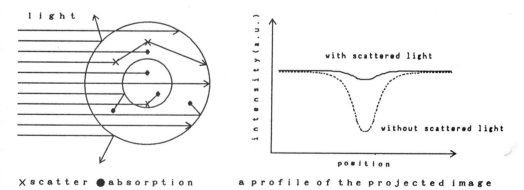

Fig. 7   Different possible paths of photons in tissue and an intensity profile of the projected image.

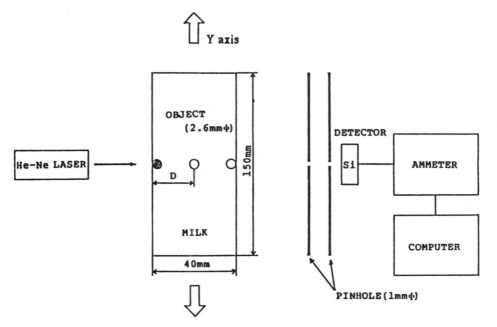

Fig. 8 Schematic diagram of the scanning system.

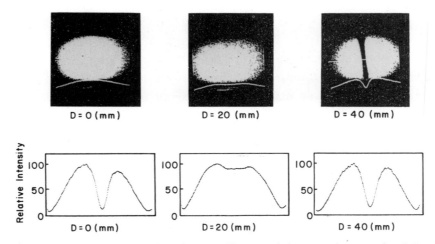

D = 0 (mm)   D = 20 (mm)   D = 40 (mm)

D = 0 (mm)   D = 20 (mm)   D = 40 (mm)

Fig. 9 Transilluminated images of the phantom. The top pictures are images of a 2.6mm wire at different depths in fresh milk obtained with a broad light source and an SIT camera. Intensity profiles are also recorded on the images. The lower pictures show the intensity profiles as obtained by the scanning system.

under normal circumstances, the object is unresolvable until almost at the front (detector) face of the phantom. Considerable improvement may be achieved by scanning a collimated detector synchronously with a laser[16,17]. Modulation is high when the object is placed near the edges of the phantom, but is some what worse when in the center, where the combined effects of the diffused input beam and large PSF become maximized.

When looking at attenuation as a function of wavelength, each tissue has its own characteristics value. The attenuation of combined red and NIR light in various tissues of a fixed path length is reported as follows[18]:

Least attenuation: Cyst
                       Fat
                       Glandular
                       Solid Tumors

Most attenuation: Blood

These differences in attenuation result in different relative intensities of transmitted light.

In our system, two sets of data are obtained from one scan: one representing the transilluminated image with NIR light (830nm), and the other illuminated with red light (630nm). The two transmission values are calculated by computer, and the results show a specific attenuation value corresponding to tissue characteristics.

Outline of the system

A schematic diagram of the system is shown in Figure 10. Inside the system, laser beams are guided through optical fiber collimators (NIPPON SHEET GLASS CO., LTD.), and face receiving fiber collimators. The transmitting and receiving fibers face each other, and are moved along two horizontal axes by two stepping motors controlled by a personal computer. The maximum dimensions of the scanned area are $18 \times 15 cm^2$ and each pixel size is $1 \times 1 mm^2$. The time required for scanning is dependent on the size of the area to be scanned, but is typically around 90 seconds.

Transmitted light signals (red and NIR) are assigned digital values using analogue to digital converters and stored in respective computer memories. The two sets of data are then averaged, and the results are used to create a pseudocolor display image. Images may be generated as necessary using the differences between red and NIR transmissions.

Clinical applications

The system is currently undergoing trials at Hamamatsu University School of Medicine. The breast is placed between the light source and detector side windows and

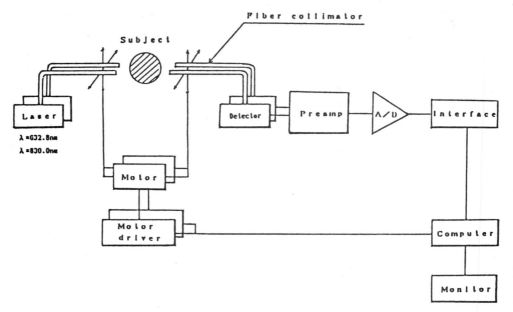

Fig. 10. Schematic diagram of the breast cancer imaging system.

slightly compressed to a thickness of 4-6 cm. Lateral scanning may be achieved by rotating the detector head by ±90º from the axial scanning position.

The image of a normal breast is shown in Figure 11[19]. Red pixels indicate a high transmission rate, and blue pixels indicate a low transmission rate. The outer (dark red) section represents fatty tissue, and the internal (blue) section represents the core consisting of glandular tissue. This image is relatively uniform.

In an abnormal case with benign mass, X-ray mammography and transillumination were performed for the comparative study. Figure 12 shows the X-ray mammography

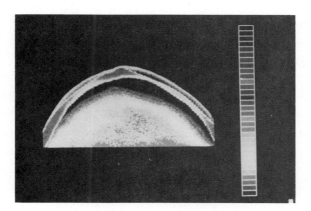

Fig. 11. Image of a normal breast observed with the system.

photograph, and a circular shadow of about 1cm diameter corresponding to the benign mass may be seen. In the transillumination image obtained using a wavelength of 830nm (Figure13), the region around the mass is shown as a higher bright area, compared with the surrounding normal tissue, while tumors show higher absorption.

Our system is still in the early stage of development. Continued evaluations and studies are necessary before its definitive utility in diagnosis of breast cancers can be determined.

CONCLUSION

As demonstrated above, it is possible to obtain physiological and anatomical information using light in the human body. The system for the cerebral oxygenation

Fig. 12. X-ray mammogram showing a benign mass in the left lower breast.

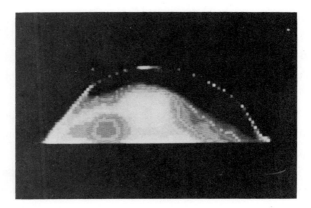

Fig. 13 Corresponding transilluminated image showing the benign mass as an area of lower absorption.

monitoring has a capability of measuring absorption changes of 0.02 O.D. against a background attenuation of 10 O.D. with a drift rate of less than 0.004 O.D./hr. Applying the average optical path length in the brain calculated with respect to water absorption[6], quantitative measurement of oxygen changes is possible.

In the breast cancer imaging system, our experiments showed that the system can detect a small tumor of less than 1cm in size. However, sufficient data are not available to document the ability of the system to detect preclinical tumors of the breast. Additional data are currently being collected to determine this. At the present time, we regard the system as being a valuable adjunct to physical examination and mammography in evaluation of diseases of the breast.

We believe that by concentrating these techniques and knowledge, we stand a good chance of developing new photonics technologies, such as an optical CT system, contributing further to the future of medicine.

ACKNOWLEDGEMENTS

We are grateful to Dr. Delpy and his colleagues of University College London for advice. We would also like to express our gratitude to Dr.Kaneko and Ms. Hè Ping of Hamamatsu University School of Medicine for useful discussions and for offering the clinical data in the breast cancer imaging system.

REFERENCES

1. F.F. Jöbsis, Noninvasive, infrared monitoring of cerebral and myocardial oxygen sufficiency and circulatory parameters, *Science*, 198:1264-1267 (1977).
2. J.S. Wyatt, M. Cope, D.T. Delpy and S. Wray, Quantification of cerebral oxygenation and haemodynamics in sick newborn infants by near infrared spectroscopy, *Lancet*, 8515:1063-1066 (1986).
3. O. Hazeki and M. Tamura, Quantitative analysis of hemoglobin oxygenation state of rat brain in situ by near-infrared spectrophotometry, *J. Appl. Physiol.*, 64:796-802 (1988).
4. M. Cope and D.T. Delpy, A system for long term measurement of cerebral blood and tissue oxygenation on newborn infants by near infrared transillumination, *Med. Biol. Eng. Comput.*, (in press).
5. Cope, M., D.T. Delpy, E.O.R. Reynolds, S. Wray, J. Wyatt and P. van der Zee, in: "Oxygen Transport to Tissue, X," M. Mochizuki, et al., eds., Plenum, New York, (in press).
6. S. Wray, M. Cope, D.T. Delpy, J.S. Wyatt and E.O.R. Reynolds, Methods of quantitating cerebral near infrared spectroscopy data, *Biochem. Biophys. Acta*, (in press).

7. R.H. Gold, E.A. Sickles, L.W. Bassett, M. McSwceney, S.A. Feig and J.R. Milbrath, Diagnostic imaging of the breast, *Invest. Radiol.*, 19:S43-S53 (1984).

8. M. Ohlsson, B. J. Gundersen and D.-M. Nilsson, Diaphanography: A method for evaluation of the female breast, *World J. Surg.*, 4:701-707.

9. C.R.B. Merritt, A. Sullivan, A. Segaloff and W.P. McKinnon, Real-time transillumination light scanning of the breast, *Radio Graphics*, 4:989-1009 (1984).

10. V. Marshall, D.C. Williams and K.D. Smith, Diaphanography as a means of detecting breast cancer, *Radiology*, 150:339-343 (1984).

11. W.A.G. Bruls and J.C. van der Leun, Forward scattering properties of human epidermal layers, *Photochem. Photobiol.*, 40:231-242 (1984).

12. J.M. Maarek, G. Jarry, B. de Cosnac, A. Lansiart and Bui-Mong-Hung, A simulation method for the study of laser transillumination of biolobical tissues, *Ann. Biomed. Eng.*, 12:281-304 (1984).

13. C.C. Johnson, Optical diffusion in blood, *IEEE Trans. Biomed. Eng.*, BME-17:129-133 (1970).

14. C.J. D'Orsi, R.J. Bartrum and M.M. Moskowitz, Lightscanning of the breast, in: "Breast Cancer Detection," L.W. Bassett and R.H. Gold, eds., Grune & Stratton, Orlando (1987).

15. M. Kaneko, S. Goto, T. Fukaya, H. Isoda, T. Hayashi, T. Hayakawa and Y. Yamashita, Fundamental studies of image diagnosis by visual lights, *Medical Imaging Technology (JAMIT)*, 2S:83-84 (1984).

16. G. Jarry, S. Ghesquiere, J.M. Maarek, F. Fraysse, S. Debray, Bui-Mog-Hung. D. Laurent, Imaging mammalian tissues and organs using laser collimated transillumination, *J. Biomed. Eng.*, 6:70-74 (1984).

17. P.C. Jackson, P.H. Stevens, J.H. Smith, D. Kear, H. Key and P.N.T. Wells, The development of a system for transillumination computed tomography, *British J. Radiol.*, 60:375-380 (1987).

18. B. Waxman, Transillumination light scanning for the diagnosis of breast cancer, Health Technology Assessment Report, National Center for Health Services Research, Rockville, NCHSR 84-176 (1984).

19. M. Kaneko, Hè Ping and T. Nishimura, Breast cancer diagnosis by transmission laser photo-scanning, in: "VI European Congress of Radiology," Lisbon, Abstracts Book, p. 208 (1987).

DEVELOPMENT AND VALIDATION OF MULTIWAVELENGTH ALGORITHMS FOR IN VIVO NEAR INFRARED SPECTROSCOPY

Claude A. Piantadosi and Benjamin J. Comfort

Department of Medicine
Duke University Medical Center
Durham, North Carolina 27710

INTRODUCTION

The ability to measure changes in tissue oxygenation noninvasively is of potential benefit to managing a variety of problems in clinical medicine. One useful way to assess tissue oxygenation in situ is to measure oxygen-dependent light absorption by hemoglobin and other chromophores present in the tissues. Soft tissues and bone are relatively translucent to near infrared (NIR) photons and three important biological molecules, hemoglobin ($Hb+HbO_2$), myoglobin ($Mb+MbO_2$) and oxidized cytochrome $\underline{c}$ oxidase (cyt $\underline{a},\underline{a_3}$) have oxygen-dependent absorption spectra in the 700-900 nm wavelength region. Therefore, NIR spectroscopy offers a potential solution to the problem of noninvasive measurement of changes in tissue oxygenation. These concepts were first reported from Duke University in 1977 when Jöbsis[1] demonstrated a "window" for transmission of NIR light in biological tissues and the feasibility of in vivo NIR monitoring of changes in the oxygenation state of hemoglobin and the oxidation level of cyt $\underline{a},\underline{a_3}$ in the brain. Continuous, noninvasive NIR monitoring techniques to evaluate tissue oxygenation have evolved gradually over the past 10 years at Duke and other institutions[2-4].

Successful interpretation of NIR signals acquired from biological tissues depends foremost upon the ability to distinguish changes in absorption resulting from simultaneous changes in the relative concentrations of the intervening absorbers. The broad NIR absorption spectra of Hb, HbO_2 and oxidized cyt $\underline{a},\underline{a_3}$ overlap extensively, thus absorption changes at a single wavelength may reflect changes in O_2-dependent absorption by either hemoglobin or oxidized cyt $\underline{a},\underline{a_3}$. Fortunately, most other O_2-dependent biomolecules either do not absorb NIR light significantly or are not present in large amounts in mammalian tissues. This generalization includes reduced cyt a,a_3 because the copper atom(s) of the enzyme complex that contribute the NIR absorption band only absorb significantly when present in the oxidized valence state[5]. In tissues containing myoglobin, the spectral analysis is not appreciably more difficult because of the near coincidence of the NIR spectra of MbO_2 with HbO_2 and Mb with Hb. The presence of myoglobin complicates physiological interpretation of NIR signals acquired from muscle tissues, however, absorption by myoglobin may be accounted for qualitatively in a straightforward manner by summing it to absorption by hemoglobin, as discussed later.

Independent NIR measurements of the relative amounts of HbO_2, Hb, and cyt a,a_3, in tissue (t) requires that contributions by the other absorbers be subtracted from the spectral response of the molecule of interest. The original dual wavelength techniques[6] obtain difference signals by subtracting optical density changes at two nearby wavelengths. This corrects for equal light scattering and absorption changes at the two wavelengths, but does not compensate for absorption changes by other absorbers that contribute unequally to the two wavelengths. Such problems can be corrected by applying algorithms that calculate values for $tHbO_2$, tHb and cyt a,a_3 by weighting changes in absorption at several wavelengths according to the relative contribution of each absorber at each wavelength. Since three overlapping absorption spectra must be deconvoluted, absorption data are needed from at least three NIR wavelengths to account for the contributions of the three molecular species. In some situations, such as those encountered with the use of reflectance geometry, four wavelength algorithms have been found empirically to provide more accurate approximations of in situ absorption properties. A brief overview of the development of multiwavelength NIR algorithms has been presented by Jöbsis[7]. This approach, sometimes identified by the acronym NIROS-SCOPY, has been applied in various laboratory and clinical settings to evaluate dynamic trends in tissue oxygenation.

NIR multiwavelength algorithms are algebraic expressions that depend upon relative absorption coefficients derived by empirical techniques. These techniques, however different, should fulfill the following general criteria: 1- derive the relevant spectra independently, i.e. without spectral contributions from other O_2 dependent molecules, 2- measure relative extinction values at a given wavelength under identical conditions of light scattering and pathlength for each absorber and, 3- acquire spectra at optical geometries and scattering functions similar to the actual biological monitoring application. This paper describes techniques to derive NIR multiwavelength algorithms based entirely upon in vivo data and fulfilling, insofar as possible, the three criteria above. In addition, instrumentation developed for continuous NIR data acquisition and deconvolution of the spectra has been described briefly and a synopsis included of recent validation and applications studies using the new algorithms.

Multiwavelength algorithms relate changes in the relative quantities of $tHbO_2$, tHb and cyt a,a_3 to each other according to total absorption changes at each of three or more wavelengths. Minor differences in wavelength-dependent light scatter and pathlength and the absence of reliable measurements of optical pathlength under in vivo conditions, prevents NIR multiwavelength data from being expressed either as changes in concentration or as true changes in optical density (OD). Optical density has been substituted for by the term variation in density (vd) to indicate that weighting coefficients derived at different wavelengths have been used to deconvolute the spectra. Variations in density are expressed in logarithmic form according to the Beer-Lambert relation, thus, the units are proportional to concentration. One vd unit has been defined as a 10-fold change in a signal computed through the appropriate algorithm[7]. The biophysical constraints above relegate the continuous NIR technique to trend monitoring at present. However, O_2-dependent changes in the relative amounts of $tHbO_2$, tHb or cyt a,a_3 accompanying physiological perturbations can be compared between experiments despite minor differences in geometry between incident radiation and photodetector. Small differences in geometry contribute to the statistical variance of the absolute data, but this variability may be lessened by expressing the values as fractions of the total NIR signal acquired under two discrete and reproducible physiological conditions.

METHODS

In Vivo NIR Spectra

NIR absorption spectra of $tHbO_2$, tHb and oxidized cyt a,a_3 have been obtained independently from the in situ brains of anesthetized laboratory animals. This has been accomplished in small cats and rats by exchange transfusing them extensively with fluorocarbon emulsions at hyperbaric pressures of oxygen as described below. Similar blood substitution experiments in the hindlimb of the rat have been performed to confirm the shapes and relative intensities of the spectra of $tHbO_2$, tHb and cyanide-inhibited cyt a,a_3 in a tissue containing myoglobin. Only results of brain experiments have been described here for the sake of simplicity and ease of interpretation.

Adult cats of either sex, anesthetized with pentobarbital (38 mg/kg) were used in experiments to derive NIR algorithms. The animal experiments were conducted within a large multipurpose hyperbaric chamber. The cats were prepared surgically with tracheostomy, arterial and venous femoral catheters and splenectomy. The carotid arteries were isolated and ligatures placed loosely around them. The animals were paralyzed and ventilated mechanically to prevent artifacts from respiratory motion. The arterial CO_2 tension ($PaCO_2$) was adjusted to approximately 30 Torr. The hair was removed from the scalp and two optical fiber bundles were pressed tightly against the cranium through the intact skin using the reflectance mode. The reflectance arrangement provided a center-to-center distance of four cm between the two optical fiber bundles. The optical path bridged both cerebral hemispheres across the sagittal suture. The substantial distance between the fiber bundles minimized detection of back-scattered incident light from superficial structures such as the cranium that would increase the ratio of scattered to transmitted light measured by the photodetector. Back-scattered light contains little absorption data and adversely affects the quality of the spectra of the deeper cerebrocortical tissue.

Incident light from an incandescent lamp was delivered to the brain by one of the fiber bundles after it passed through a holographic grating monochromator (JY H-10 VIS-IR) and a slotted chopping wheel. The spectral bandwidth was 6 nm. Light diffusely reflected from the brain was collected by the second fiber bundle and measured with a photomultiplier (Hamamatsu R936). Signals at each wavelength were measured by integrating the anode current using an operational amplifier and subtracting the dark current signal. Ambient light was excluded by taping the area around the fiber bundles with black tape and draping the preparation with black cloth. All spectra were recorded in 5 nm intervals from 720 to 920 nm.

After preparation, the animals were infused constantly with pentobarbital at 8 mg/kg/hr and ventilated on 95% O_2 and 5% CO_2. The pressure in the hyperbaric chamber was increased to 4 ATA in order to arterialize the venous blood returning to the vena cavae. Baseline spectra were recorded and the animals were exchange transfused with 8 blood volumes of perfluorotributylamine emulsion (FC-43, Alpha Therapeutics, Los Angeles). Spectra were repeated after 2,4,6, and 8 blood volumes of exchange. These spectra represent in vivo HbO_2 spectra because the venous hemoglobin saturation was not allowed to fall below 100%. The HbO_2 spectra were recorded while the oxygen content of the arterial fluorocarbon emulsion was supraphysiologic, i.e. approximately 36 ml/dl, at the end of the exchange transfusion. This enabled us to keep the copper band of cyt a,a_3 fully oxidized throughout the exchange transfusion, thus preventing it from distorting the HbO_2 spectra.

After the HbO_2 had been replaced with FC-43, the heart was fibrillated with KCl and after 15 minutes, another spectrum recorded. The difference between the brain spectrum after death and the one immediately after completing the exchange procedure yielded the difference spectrum of the NIR band of oxidized cytochrome a,a_3. The final spectrum (tHb) was obtained after death by cannulating both carotid arteries from beneath (without moving the head or optical bundles) and infusing deoxygenated red blood cells (RBC) into the brain. The blood removed during the exchange was centrifuged, washed and the RBC packed to a hematocrit of 70-80%. The RBC suspension was deoxygenated with crystalline $Na_2S_2O_4$ and pumped into the brain via both carotids at a pressure of about 100 mm Hg. A final spectrum was recorded and the previous spectrum subtracted from it to furnish a spectrum of tHb.

Continuous NIR Spectroscopy

NIR spectroscopy was performed using a laser-based instrument designed and constructed in our laboratory for simultaneous interrogation of two tissues at three or four wavelengths each. A brief description of the instrument has been published[8]. Monochromatic light (1.5nm bandwidth) illuminates the tissues from a bank of GaAlAs laser diodes at 775, 810, 870 and 904 nm. Using time-domain multiplexing, the laser diodes are pulsed at a frequency of 1KHz and a pulse width of 200 ns. The incident light intensity is monitored with a reference (R) photodiode and the strength of the signals (S) from the tissue measured by a photomultiplier. The photocurrents from the two photodetectors are integrated, demultiplexed and fed through a log ratio amplifier. The log S/R voltages are then input to a dedicated microprocessor where the algorithms are applied. The $tHbO_2$, tHb and cyt $a,a,3$ signals are displayed in real time on a printer and the log S/R values recorded on disc. A simple diagram of the instrument is provided in Figure 1. At initial conditions, the signals are adjusted to S=R at each wavelength, so that -log S/R=0. We assume that Δ S/R is independent of wavelength at the extended pathlengths employed to monitor tissue in the NIR region.

RESULTS AND DISCUSSION

NIR Algorithms

The $tHbO_2$ and tHb spectra were scaled to equimolar quantities using apparent _in vivo_ extinction coefficients and the total ΔOD measured at the 810 nm hemoglobin isosbestic point during the FC-43 exchange transfusion. Although normal tissue contains more oxyhemoglobin than deoxyhemoglobin, equalizing the initial concentrations of $tHbO_2$ and tHb assures that any change in the amount of one compound will be offset exactly by an opposite

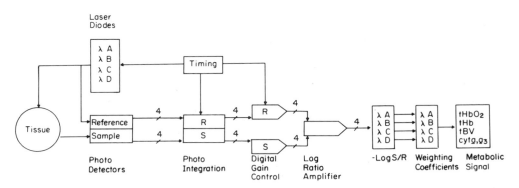

Fig 1. Schematic diagram of NIR spectrophotometer

change in the amount of the other. The ΔOD at the hemoglobin isosbestic point and the maximum ΔOD measured for copper band of cyt a,a3 were used to compute a hemoglobin to cyt a,a3 ratio for the algorithm derivation. Representative NIR spectra of tHbO2, tHb and oxidized cyt a,a3 from a single cat are provided in Figure 2. A more detailed analysis of the absorption characteristics of the cyt a,a3 copper band acquired in similar experiments and procedures for stripping residual traces of hemoglobin from the spectrum of the copper complex have been published recently[9].

Algorithms were then derived for three wavelengths using the following linear relationships:

$$\Delta OD_{775nm} = a\Delta OD(HbO_2)_{775nm} + b\Delta OD(Hb)_{775nm} + c\Delta OD(cyt\ a,a_3)_{775nm}$$

$$\Delta OD_{810nm} = d\Delta OD(HbO_2)_{810nm} + e\Delta OD(Hb)_{810nm} + f\Delta OD(cyt\ a,a_3)_{810nm}$$

$$\Delta OD_{904nm} = g\Delta OD(HbO_2)_{904nm} + h\Delta OD(Hb)_{904nm} + i\Delta OD(cyt\ a,a_3)_{904nm}$$

where a, b, c, f, g, h and i are fractional absorption values relative to values of 1.0 for d and e. These three linear expressions were solved simultaneously by matrix inversion to obtain three new equations:

$$\Delta OD\ Hb\ = a^{-1}\Delta OD_{775nm} + d^{-1}\Delta OD_{810nm} + g^{-1}\Delta OD_{904nm}$$

$$\Delta OD\ HbO_2 = b^{-1}\Delta OD_{775nm} + e^{-1}\Delta OD_{810nm} + h^{-1}\Delta OD_{904nm}$$

$$\Delta OD\ cyt\ = c^{-1}\Delta OD_{775nm} + f^{-1}\Delta OD_{810nm} + i^{-1}\Delta OD_{904nm}$$

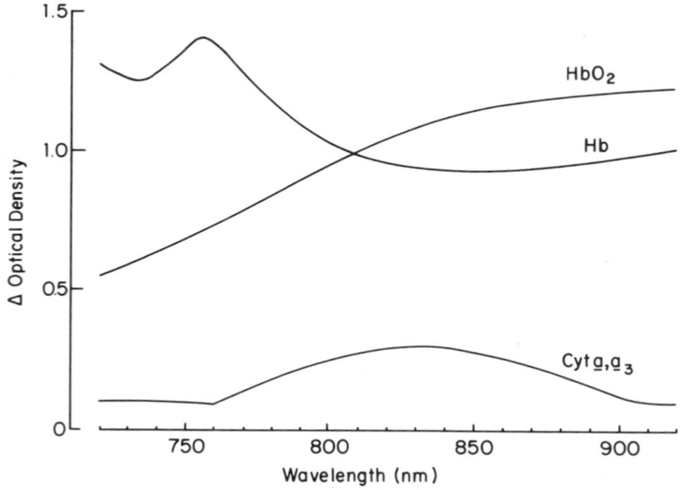

Fig 2. In Vivo spectra of tHb, tHbO2 and oxidized cyt a,a3 acquired from the brain of an anesthetized cat.

where a^{-1} through i^{-1} are the weighting coefficients from the inverse matrix. A set of nine weighting coefficients were derived for each of the eight sets of spectra and the values at each wavelength averaged to obtain general coefficients.

Four wavelength algorithms also were derived from the same sets of spectra using 870 nm as the fourth wavelength. In this situation, the fourth linear equation may be used either to define a light scattering function or to correct for a small irregularity (fo) of uncertain origin on the long wavelength side of the 830 nm copper band of cyt a,a3. Thus, the initial equations are set-up as:

$$\Delta OD_{775nm} = a\Delta OD(HbO_2) + b\Delta OD(Hb) + c\Delta OD(cyt\underline{a},a3) + d\Delta OD \ (fo)$$

$$\Delta OD_{810nm} = e\Delta OD(HbO_2) + f\Delta OD(Hb) + g\Delta OD(cyt\underline{a},a3) + h\Delta OD \ (fo)$$

$$\Delta OD_{870nm} = i\Delta OD(HbO_2) + j\Delta OD(Hb) + k\Delta OD(cyt\underline{a},a3) + l\Delta OD \ (fo)$$

$$\Delta OD_{904nm} = m\Delta OD(HbO_2) + n\Delta OD(Hb) + o\Delta OD(cyt\underline{a},a3) + p\Delta OD \ (fo)$$

A correction to the 830 nm band was derived empirically using Lorentzian analysis of the line shapes as previously reported[10]. The results of the curve fitting indicated that two functions described the composite band adequately. The dominant component had a wavelength maximum of approximately 820 nm while the minor component had a wavelength maximum at approximately 870 nm. Since no definite identity has been assigned to the smaller band, it has been subtracted from the larger band in the four wavelength algorithms. The fourth algorithm thus obtained is not used in routine in vivo NIR monitoring at the present time.

Validation of NIR Algorithms

Several series of validation experiments have been performed to evaluate how well the three and four wavelength algorithms independently measure changes in $tHbO_2$, tHb and cyt a,a3. A useful validation experiment is to repeat the fluorocarbon-for-blood substitution under direct NIR observation. NIR spectroscopy of the brain has been performed directly through skin and skull during the process of FC-43 for blood exchange transfusion in both cats and rats anesthetized with pentobarbital and mechanically ventilated to maintain constant $PaCO_2$.

The results of a representative experiment from a cat are shown in Figure 3. In this experiment, three wavelength algorithms were used in the reflectance mode to monitor changes in cerebrocortical $tHbO_2$, tBV (blood volume measured as total hemoglobin, $tHbO_2$ + tHb) and cyt a,a3 oxidation state during FC-43 exchange transfusion. The animal

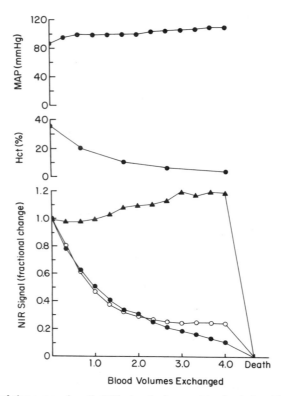

Fig. 3. Responses of three wavelength NIR signals from cat brain during FC-43-for-blood exchange transfusion. Top panel shows accompanying changes in mean arterial blood pressure and hematocrit. Bottom panel shows responses of cytochrome (triangles), tHbO2 (open circles) and tBV (closed circles).

breathed 100% O2 during the experiment and the electroencephalogram (EEG) activity was well maintained throughout the period of the exchange. The disappearance of hemoglobin was exponential in accordance with the change in hematocrit (hct). There was little change in the cyt a,a3 oxidation level early in the exchange, indicating effective separation of absorption by tHbO2 from the oxidized copper by the cytochrome algorithm. After four blood volumes of exchange transfusion (hct 2%), the heart was arrested and the cyt a,a3 copper band disappeared with brain ischemia. There was little change in tHbO2 and tBV because there was only a small amount of remaining hemoglobin. These responses confirmed effective separation of the hemoglobin and cyt a,a3 absorption spectra.

Figure 4 shows composite results of similar hemoglobin washout experiments in 11 rats. In these experiments, the brain was monitored using four wavelength NIR algorithms in the transillumination mode. The animals breathed 100% O2 during the exchange process. Once again, no statistically significant change in cyt a,a3 oxidation level was

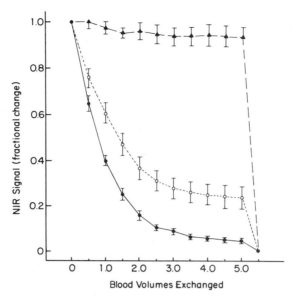

Fig. 4. Changes in brain tBV, tHbO2 and oxidized cytochrome during FC-43-for-blood exchange transfusion in the rat. Symbols are the same as Figure 3. Values are mean ± SE for n=11.

detected by the algorithm for the copper band until the circulation was arrested. At this point, the copper band disappeared with small changes in the tHbO2 and tBV signals. The results of these experiments indicate a maximum error in determination of cyt a,a3 of 10%, assuming that changes in cyt a,a3 calculated by the algorithm during the washout were due only to crosstalk from tHbO2. This assumption, however, overestimates the amount of crosstalk because changes in cyt a,a3 oxidation level usually occurred after most of the tHbO2 had been eliminated. The actual error signal is being determined currently by statistical analyses.

A similar series of hemoglobin washout experiments have been conducted in the hindlimb of the rat. Again, the separation of HbO2 and tBV from cyt a,a3 in the NIR can be demonstrated during hemoglobin elimination. Independent proof that myoglobin does not contribute to the cyt a,a3 algorithm has been difficult, however, the NIR spectrum of myoglobin is so similar to hemoglobin that separation of the cytochrome from hemoglobin also validates the independence of cytochrome and myoglobin signals. Myoglobin is most easily accounted for by using the HbO2 algorithm as a comprehensive index of the O2 store in tissues where it is present. In these tissues, changes in the tBV (tissue hemoglobin volume) may be assessed by the sum:

$$(HbO_2 + MbO_2) + (Hb + Mb)$$

if we assume that the total concentration of the two myoglobin species in the tissue is conserved for the duration of the experiment. Therefore, the only net change in the sum will result from oxy- or deoxyhemoglobin entering or leaving the optical field.

Experimental Applications

NIR multiwavelength spectroscopy has been used primarily to monitor intact brain and skeletal muscle tissues. These applications include triple wavelength studies of experimental animals during hypoxia and hypovolemia[11], elevated intracranial pressure[12], and respiratory acid-base disorders[13]. NIR observations of humans using older algorithms have included preliminary evaluation of the cerebrocortical effects of hypoxia and hyperoxia/hypercarbia in healthy volunteers, passive monitoring of adult patients under general anesthesia[14], premature newborn infants in the nursery[15,16] and aviators under high g stress[17]. The newer four wavelength algorithms derived as described here have been used to monitor brain and muscle responses to hypoxic and ischemic stress in experimental animals[18] and in human volunteers[8,19,20]. Data from some of the recent NIR monitoring studies have provided additional opportunities to test the algorithms because the NIR signals may be expected to deviate predictably from each other under the influence of certain physiological forcing functions.

Recent studies of forearm NIR responses to venous and arterial occlusion in human subjects[8] using the four wavelength algorithms have furnished changes in the NIR cytochrome signal that are independent of changes in $tHbO_2$ or tBV and vice versa. In forearm muscle, sudden ischemia is characterized by parallel decreases in $tHbO_2 + MbO_2$ and cyt a,a_3 oxidation level. This finding is illustrated by Figure 5, which displays the NIR responses to 10 minutes of complete forearm ischemia produced by inflating a blood pressure cuff to 250 mmHg. It is noteworthy that the cyt a,a_3 oxidation level in the example of Figure 5 does not fall below baseline for the initial 1.5 min of ischemia. The change in the cytochrome oxidation level clearly trails the $tHbO_2$ which begins to decrease almost immediately. This observation confirms that these NIR signals behave independently of each other. The mitochondrial oxygen uptake falls to zero after only 8 to 10 min of complete limb ischemia.

The recovery responses post-occulsion are also of interest for two reasons. The first is the noninvasive demonstration of dynamic vascular hyperemia after release of the tourniquet. The second is an oxidative overshoot by cyt a,a_3 that amounts to 50% of the total reduction response during ischemia. The oxidation may be interpreted in various ways, but it is consistent with the fact that resting skeletal muscle normally contains many closed

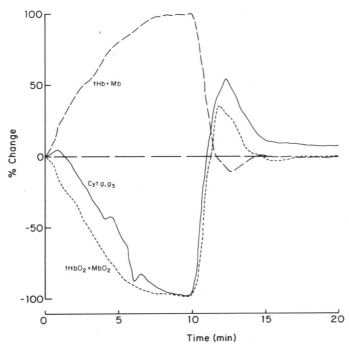

Fig. 5. NIR responses to 10 min of complete forearm ischemia and recovery produced by temporary cuff occlusion of the arterial supply.

capillaries which become perfused actively during the hyperemic response after a period of arterial occlusion.

The forearm monitoring studies have disclosed another interesting finding where $tHbO_2 + MbO_2$ and cyt a,a_3 oxidation level diverge[20]. This response was found in the resting skeletal muscle, e.g. brachioradialis, when the venous pressure in the limb approaches diastolic pressure by cuff occlusion of the upper arm (Figure 6). At low cuff inflation pressures (<60 mmHg), the tBV was shown to increase without changes in the cytochrome oxidation level or total O_2 store. As the cuff pressure approached diastolic pressure, cyt a,a_3 became reduced without a concomitant decrease in the total O_2 store ($tHbO_{2+}$ MbO_2) in the tissue. At cuff pressures above diastolic, total cytochrome reduction accompanied O_2 depletion.

A simple explanation for the divergence of the O_2 store and cytochrome signals is that the elevated venous pressure causes hypoperfusion of some capillaries and the O_2 concentration falls to zero at nearby mitochondria. The expected myoglobin desaturation is

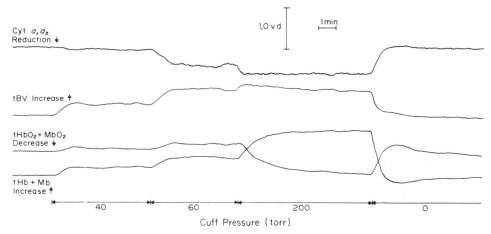

Cyt a,a_3
Reduction ↓

1,0 v.d.   1 min.

↑BV Increase ↑

↑HbO₂ + MbO₂
Decrease ↓

↑Hb + Mb
Increase ↑

Cuff Pressure (torr)
40   60   200   0

Fig. 6. NIR responses of human forearm to graded venous and arterial blood flow restriction by cuff occlusion. v.d. = variation in density

masked by a proportional increase in the amount of tHb0₂ present in adjacent small blood vessels (venoarterial reflex). The validity of the stable tHbO₂ signal in this setting has been corroborated by stable arterial oxygen saturation measured by finger pulse oximetry in the same extremity. The example of Figure 6 may be interpreted as a form of functional arteriovenous shunting in skeletal muscle resulting in a net decrease in intramitochondrial oxygen availability. Related NIR studies currently in progress may provide new opportunities for insight into mechanisms of arteriovenous shunting and venous stasis injury encountered in chronic venous insufficiency of the lower extremities.

REFERENCES

1. F.F. Jöbsis, Non-invasive, infrared monitoring of cerebral and myocardial oxygen sufficiency and circulatory parameters, *Science*, 198:1264-1267 (1977).
2. I. Giannini, M. Ferrari, A. Carpi and P. Fasella, Non-invasive near infrared spectroscopy of brain and fluorocarbon exchange transfused rats, *Physiol. Chem. Phys.*, 14:295-305 (1982).
3. O. Hazeki, A. Seiyama and M. Tamura, Near infrared spectrophotometric monitoring of hemoglobin and cytochrome a,a3 in situ, in: "Advances in Experimental Medicine and Biology," A. Silver and A. Silver, eds., Plenum Press, New York (1986).
4. J.S. Wyatt, M. Cope, D. T. Delpy, S. Wray and E. O. R. Reynolds, Quantification of cerebral oxygenation and hemodynamics in sick newborn infants by near infrared spectrophotometry, *Lancet*, 5818(2):1063-1066 (1986).

5. D.C. Wharton, and A. Tzagoloff, Studies on the electron transfer system, LVII, The Near infrared absorption band of cytochrome oxidase, *J. Biol. Chem.*, 239:2036-2041 (1964).

6. B. Chance, Rapid and sensitive spectrophotometry, III, A double beam apparatus, *Rev. Sci. Instrum.*, 22:634-638 (1951).

7. F.F. Jöbsis-VanderVliet, Non-invasive, near infrared monitoring of cellular oxygen sufficiency in vivo, in: "Advances in Experimental Medicine and Biology," F. Kreuzer, S. M. Cain, Z. Turek and T. K. Goldstick, eds., Plenum Press, New York (1985).

8. N.B. Hampson, and C. A. Piantadosi, Near infrared monitoring of human skeletal muscle oxygenation during forearm ischemia, *J. Appl. Physiol.*, 64:2449-2457 (1988).

9. F.F. Jöbsis-VanderVliet, C. A. Piantadosi, A. L. Sylvia, S. K. Lucas and H. H. Keizer, Near infrared monitoring of cerebral oxygen sufficiency, I, Spectra of cytochrome c oxidase, *Neurol. Res.*, 10:7-17 (1988).

10. H.H. Keizer, F. F. Jöbsis-VanderVliet, S. K. Lucas, C. A. Piantadosi and A. L. Sylvia, The near infrared absorption band of cytochrome a,a_3 in purified enzyme, isolated mitochondria and in the intact brain in situ, in: "Advances in Experimental Medicine and Biology," F. Kreuzer, S. M. Cain, Z. Turek and T. K. Goldstick, eds., Plenum Press, New York (1985).

11. C.A. Piantadosi, T.M. Hemstreet and F.F. Jöbsis-VanderVliet, Near infrared spectrophotometric monitoring of oxygen distribution to intact brain and skeletal muscle tissues, *Crit. Care Medicine*, 14:698-706 (1986).

12. H.J. Proctor, C. Cairns, D. Fillipo and F.F. Jöbsis-VanderVliet, Near infrared spectrophotometry: potential role during increased intracranial pressure, in: "Advances in Experimental Medicine and Biology," F. Kreuzer, S. M. Cain, Z. Turek and T. K. Goldstick, eds., Plenum Press, New York (1985).

13. N.B. Hampson, F.F. Jöbsis-VanderVliet and C.A. Piantadosi, Skeletal muscle oxygen availability during respiratory acid-base disturbances in cats, *Respir. Physiol.*, 70:143-158 (1987).

14. F.F. Jöbsis-VanderVliet, E. Fox and K. Sugioka, Monitoring of cerebral oxygenation and cytochrome a,a_3 redox state, in: "International Anesthesiology Clinics," K.K. Tremper, ed., Little, Brown and Co., Boston (1987).

15. J.E. Brazy, D.V. Lewis, M.H. Mitnick and F.F. Jöbsis-VanderVliet, Non-invasive monitoring of cerebral oxygenation in pre-term infants: preliminary observations, *Pediatrics*, 75:217-225 (1985).

16. J.E. Brazy and D. V. Lewis, Changes in cerebral blood volume in cytochrome $a,a3$ during hypertensive peaks in pre-term infants, *J. Pediatrics*, 108:983-987 (1986).

17. D.H. Glaister, Current and emerging technology in G-LOC detection: Non-invasive monitoring of cerebral microcirculation using near infrared, *Aviat., Space, Environ. Med.*, 59:23-28 (1988).

18. J.A. Griebel, N.B. Hampson and C.A. Piantadosi, Cytochrome redox states correlated with O_2 uptake and oleic acid lung injury in rabbits, *Am. Rev. Resp. Dis.*, 137:115A (1988).

19. N.B. Hampson, E.M. Camporesi, R.E. Moon, B.W. Stolp, J.A. Griebel, S.L. Whitney and C.A. Piantadosi, Effects of hypocapnic and normocapnic hypoxia on cerebral oxygenation and ventilatory responses in humans, *Am. Rev. Resp. Dis.*, 137:143A (1988)

20. C.A. Piantadosi, J.A. Griebel and N.B. Hampson, Intramitochondrial oxygenation decreases in forearm muscle during venous congestion, *Clin. Res.*, 36:373A (1988).

APPROACHES TO THE QUANTIFICATION OF TISSUE PIGMENTS BY NIR IN VIVO USING A DOUBLE BEAM METHOD TO MULTICOMPONENT CURVE FITTING ANALYSIS

[a]Shoko Nioka and [b]Kouich Oka

[a]Dept. of Biochemistry/Biophysics
University of Pennsylvania
Philadelphia, PA

[b]Otsuka Electronics, Ltd.
Osaka, Japan

INTRODUCTION

Near infrared spectroscopy (NIR) can be used as a non-invasive method for measuring tissue oxygen transport and metabolism by indicating the levels of hemoglobin saturation and the redox state of cytochrome oxidase. Jobsis and others[1,11] have applied NIR spectroscopy to isolated mitochondria, and tissues, *in vitro* and *in vivo*. One of the advantages of using near infrared wavelengths rather than visible light is that the penetration of the photons is much greater because of the physical properties of light and the low extinction coefficients of the tissue pigments at these wavelengths. It has been found that the mean light path length varies with wavelength when measured by time resolved spectroscopy (TRS)[2] and other techniques[3]. Physical studies have indicated that in a scattering medium, such as brain tissue and cells, the photons migrate as a diffusing or randomly walking particle making application of the Beer-Lambert law difficult[4,5,6]. The light pathlength varies depending upon the wavelengths, materials, and concentrations of absorbing pigments. Theoretical equations have been developed to fit the data[7,8]. Quantification of the scattering factor is necessary in order to quantitate tissue concentration of absorbers. The contribution of the absorber to the total absorption in the scattering medium depends upon the wavelength as well as the extinction coefficient[9,5]. Time resolved spectroscopy shows similar properties of light diffusion (distribution of the light path) at 700 to 800 nm (Nioka, unpublished data). Near infrared light may be best suited to minimizing the scattering problem. Some researchers have ignored the effect of scattering and calculated the absorptions of hemoglobin and cytochrome oxidase separately[10-15,1].

This paper describes attempts to a) calibrate *in vivo* NIR spectroscopy using a yeast-hemoglobin and calcium carbonate-hemoglobin model; b) calculate the light path at different absorber concentrations; c) evaluate the ability of dual beam spectroscopy to quantitate hemoglobin saturation; and d) introduce a new approach to the measurement of the redox state of cytochrome oxidase by multicomponent curve fitting analysis.

METHODS

Path length Study of Interactive Remission and Transmittance in the Yeast Hemoglobin Mixture

Yeast concentrations of 5.6% and 8.6% (dry/water) were used; the cell volume ratio were 17% and 26% in those solutions respectively. 10 μM to 1 mM of human hemoglobin was added to the yeast and absorption spectra from 600 to 1100 nm were taken using a photodiode array (Otsuka Electronics). Transmittance and interactive light propagation were measured in the same solution with the same light source and detector probe separated by 1 cm. When measuring reflectance, the two light guides were angled at 55° and separated by 1 cm to mimic *in vivo* experiments on the skull.

Hemoglobin and Cytochrome c Oxidase Spectra in a Scattering Medium

The distortion of spectra resulting from tissue scattering was investigated by comparing those in a crystalloid solution and in a scattering medium such as yeast and $CaCO_3$, both containing hemoglobin or cytochrome c oxidase (cytochrome aa_3). The purpose of this study was to determine the extinction coefficient as a function of scattering. In the animal model brain tissue spectra under erythrocyte-free conditions were compared to *in vitro* models.

Correlation of Hemoglobin Saturation in Brain Tissue to the Sagittal Sinus Blood Using Dual Beam Spectroscopy

A Johnson Foundation dual wavelength air turbine spectrophotometer was used at 850 nm and 782 nm to illuminate a 3 m long, 3 mm diameter fiber optics bundle. The illuminated light guide was located on the skull of a dog and interactive light was detected with an identical light guide situated 1 cm away at a 50-60° angle to the emerging light guide. The emergent light was coupled to a photomultiplier (Hammamatsu R928). The two wavelengths were time shared each at 8 ms. A J-Y type monochromator with curved holographic gratings blazed for the near infrared region was illuminated by a 59 W halogen lamp through a 1 mm exit slit. Hemoglobin saturation was calculated as follows: the maximum available hemoglobin saturation signal (OD 782-850) during a recovery time from hypoxia was used as 100% HbO_2, and 0% was taken at death.

Component Analysis of Brain Hemoglobin and Cyt aa_3 by Curve Fitting

Multiwavelength spectra were taken with an Otsuka MCLD 100 photodiode array system. White light (120 watt tungsten light source) carried through a 50 cm long, 3 mm quartz light guide, was shone on the skull. The detector light guide was connected to a monochromator which split the light into individual photodiode cells with a 2 nm resolution. For the *in vivo* studies, it was found that the 1 cm separation between the light guides produced an adequate signal to noise ratio.

A curve fitting program based on linear minimum square routine was used to fit the raw data based on 4 extinction coefficients and quantities of four pigments (deoxy- and oxyhemoglobin, and reduced and oxidized cytochrome aa_3. Scattering factors related to extinction coefficients were taken into account by using extinction coefficients measured in a scattered medium. To reduce the complications of scattering, the analyses were initially restricted as follows: 1) the wavelength range used for the curve fitting analysis was short, 700-950 nm. In this range, the path length was not wavelength dependent. 2) Difference spectra taken between normoxia and hypoxia were analyzed. The difference spectrum was taken in a short period so that the baseline spectra corresponding to water, cells, lipid, etc. were subtracted.

Surgical Procedure

Cats, dogs, and rats were anesthetized with ketamine hydrochloride (20 mg/kg), 2% isofluorane, and 25 mg/kg pentobarbital respectively. The dogs and cats were ventilated through a tracheal tube with a Harvard animal ventilator at 0.4 Hz, and a tidal volume of 12 ml/kg to keep the $PaCO_2$ at 35 ± 5 mmHg. Arterial and venous cannulas were placed in the femoral artery and vein to deliver a continuous infusion of Ringers-bicarbonate-5% glucose solution (5 ml/kg/hr), and to monitor blood pressure. These lines were also used to analyze blood gasses intermittently. The body temperature was kept between 35-37° C by surrounding the body with water heater blankets. The animal was placed in the stereotaxic holder during the surgery and experiment. The skin and muscle overlying the top hemisphere of the skull were removed and the upper part of the skull was exposed. The sagittal sinus was cannulated through a burr hole in the middle sagittal sinus if necessary. Black plastic windows for the NIR light guides were glued to the skull at a 40° angle, 1 cm apart.

Experimental Protocol

Under normal hematocrit, hypoxia was induced by giving stepwise FiO_2 or sudden anoxia (100% N_2). Then red blood cells were exchanged by fluorocarbon emulsion (20% of perfluoro-decalin and perfluor-tripropylamine) to produce a hematocrit of less than 1% in the circulatory blood. When the expected hematocrit was obtained, hypoxia was induced stepwise to 0% FiO_2. Epinephrine maintained the MABP at 50 mmHg until the heart did not respond to 1 mg of bolus epinephrine iv. When hypoxia caused uncontrollable heart function, MABP lower than 50 mmHg, FiO_2 was turned to 0% to assure the tissue 100% deoxyhemoglobin level. When phosphorus nuclear magnetic resonance spectroscopy (P-NMR) was used to correlate % HBO_2, NMR data was collected during three 4 minute periods at each FiO_2 level while NIR spectra were continuously acquired every 30 seconds at 580 to 1000 nm. The data collection began 4 minutes after a change in FiO_2 to allow for equilibration. At an FiO_2 of 9% the PCr/Pi (creatine phosphate/inorganic phosphate ratio) as determined by NMR began to drop. At the first sign of a PCr/Pi reduction, further decrements in FiO_2 were made in 1%-0.5% steps and each level of FiO_2 was maintained for 16 min. Since PCr/Pi was unstable at 1 or lower, the FiO_2 was adjusted to maintain the same PCr/Pi level. We found that it was impossible to maintain a PCr/Pi lower than 0.5 (in all age groups) because the energy state was too unstable which could cause cardiac failure.

RESULTS AND DISCUSSION

Measurement of Light Path Length

Reflectance and transmittance data of yeast and cat brain homogenates measured at 600 to 1100 nm were used to calculate the path length. The transition of oxyhemoglobin to deoxyhemoglobin was observed as the yeast depleted the dissolved oxygen 5 min. after the oxygen supply was stopped. The Beer-lambert law was used to calculate the apparent average light path length from the measured absorption, the known extinction coefficient, and the hemoglobin concentration in the yeast. The results, shown in Figure 1 are plotted as a function of hemoglobin and yeast concentrations. As the yeast concentration was increased from 5.6% to 8.4%, the average path length doubled. The hemoglobin concentration altered the average light path length and had an inverse relation to the light path length. As the hemoglobin concentration became higher, the average light path length decreased. This phenomenon can be easily understood as the probability of a photon being absorbed is related to the length of migration, the longer the distance the greater the probability of absorption. Consequently, the average light path becomes shorter with higher hemoglobin concentration. The cat brain homogenate with a physiological hemoglobin concentration had approximately 4 cm average light path length, 4 times longer than the distance of the two light guides. Homogenated cat brain showed the same phenomena of light path changes according to hemoglobin concentration. This phenomenon was observed with the same set of studies using dual beam spectroscopy.

Hemoglobin and Cyt *aa3* in a Scattering Medium

The absorption of hemoglobin and cytochrome *aa3* measured in a scattering medium are shown in Figures 2 and 3. In Figure 2, the absorption of deoxyhemoglobin in a scattering medium is different from that in a crystalloid solution at 600-700 nm; they exhibit similar absorptions at 700-900 nm. Purified cytochrome *aa3* oxidase solution by itself causes scattering. Figure 3 shows spectra of purified cytochrome *aa3* concentrated and diluted 3-fold with water and 10-fold with 2% $CaCO_3$. The difference spectra of oxidized-reduced cytochrome *aa3* in the three different solutions are shown in Figure 4. The characteristics of the cytochrome *aa3* difference spectra is similar in 3 solutions, however there are differences in the ratio of absorption at 605 nm to that of 830 nm, and distinct water peaks are seen in the spectra above 900 nm.

Dual Beam Spectroscopy Evaluation of %HBO2 in Brain Tissue

The results of %HbO2 measured with the dual beam spectrophotometer exhibits a sigmoidal relationship to venous blood from the dog brains. A typical oxygen dissociation curve is shown in Figure 5 where Pv50 in this curve was 24 mmHg. During the second induced hypoxia the Pv50 shifted to the right which was attributed to the Bohr effect due to a 0.2 decrease of blood pH. The venous blood saturation correlated well in this case. It was found that hemoglobin Pv50 in 6 adult dog brains was 25.8 ±5 mm Hg compared to 26

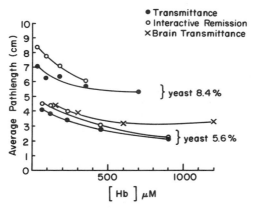

Fig. 1. Light path length (cm) as a function of hemoglobin concentration and scattering is demonstrated. The solid circle represents transmission and the open circle interactive remission. Distance between the two light guides was 1 cm in this study. Cross symbols indicate cat brain homogenate (blood-free) with transmittance light. Light path length changes inversely with concentration of the absorber (hemoglobin).

mmHg for *in vitro* measurements of dog blood. The average %HbO2 in the sagital sinus was 10% lower than that in tissue when measured spectrophotometrically. In studies using the dual beam spectrophotometer the resulting Pa50s for dogs in various age groups differed according to the level of fetal hemoglobin: new born puppies 22.2 mmHg, 2-3 week old puppies 38 mmHg, and adult dogs 40 mmHg.

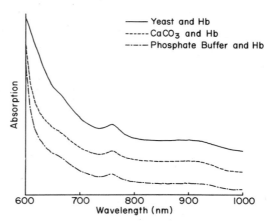

Fig. 2. Absorption spectra from 600 nm to 1000 nm of human deoxyhemoglobin ([Hb] ≈ 500μM was used) in 5.6% yeast solution (solid line), 2% CaCO3 (dotted line),and phosphate buffer solution (broken line). These spectra are similar at 700-900 nm, but differ at shorter and at larger wavelengths.

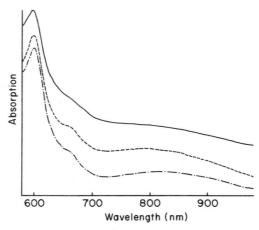

Fig. 3. Spectra of purified cytochrome aa3 (oxidized form) from 580 nm to 980 nm taken with a photodiode array (Otsuka MCPD100). Solid line; Concentrated solution (150 μM), Dotted line; diluted 3 times with phosphate buffer, and broken line; 10% diluted with 2% CaCO3. Slight differences of spectra among three different conditions are observed at all wavelengths. The scale of absorption in each spectra are different.

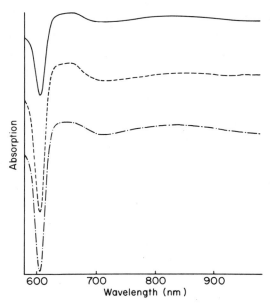

Fig. 4. Difference spectra of purified cytochrome aa_3 (oxidized form minus reduced cytochrome aa_3). Conditions of the 3 spectra are same as in Figure 3.

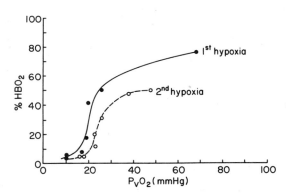

Fig. 5. The oxygen dissociation curve of brain tissue hemoglobin measured by a dual beam spectrometer at 850-782 nm. Oxygen partial pressures were measured in the sagittal sinus blood. The oxygen tension was changed by shifting the FiO_2. The Pv50 was 24 mmHg in the first hypoxic experiment. The second hypoxia yielded right shifted Pv50. The pH of the sagital sinus blood decreased from 7.4 to 7.2.

The relation of $\%HbO_2$ in brain tissue to the energy state indicator PCr/P_i as measured by NMR spectroscopy was determined for adult dogs and puppies. In adults, a 25% reduction of PCr/P_i occurred at 30% of HbO_2 and in neonates the same reduction was seen at less than 20% HbO_2. The PCr/P_i decreased to 50% of normal levels at 15% and 10% HbO_2

in adult and neonate brains respectively. These data indicate that the critical brain tissue %HbO2 is age dependent and higher in adults dogs which can be attributed to the oxygen gradient between the cytosol and the capillary bed as a function of O_2 consumption.

In vivo Brain Multiwavelength Spectra

Figure 6 presents a typical cat brain spectra taken without a skull in the 600 to 1040 nm range during normoxia (dashed line) and hypoxia (solid line). A distinct peak of deoxyhemoglobin at 760 nm and a water peak at 970 nm are readily apparent. Spectra taken at less than 620 nm are not accurate because of high tissue absorption and low penetration of short wavelengths. The peaks resulting from water and lipids at 745 nm, 840 nm, and 880 nm are seen in detail in Figure 6. Data collected when the blood was replaced by a fluorocarbon solution are shown in Figure 7. Note that absorbency changes between blood and blood-free conditions at 800 nm was about 0.2 (this difference is not shown in Figures 7 and 8).

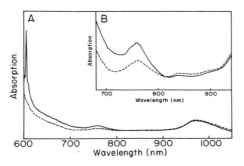

Fig. 6. A typical spectra of cat brain, with blood, using a photodiode array (Otsuka Electronics MCPD 100). The dashed line indicates normal brain spectra; the solid line represents hypoxia. There is a characteristic peak of deoxyhemoglobin at 760 nm and distinct water and lipid peaks at 980 nm, 880 nm, 840 nm, and 740 nm. Spectra lower than 620 nm were inaccurate due to high absorption.

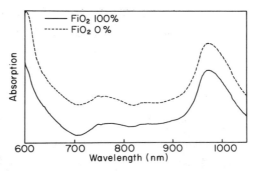

Fig. 7. A typical blood-free cat brain spectra. The blood hematocrit was less than 1% in this cat. The 760 nm deoxyhemoglobin peak remains and the water peaks are more dominant. (Solid line: 100% inspired O_2. Dashed line 0% inspired O_2).

Difference spectra of normoxia-hypoxia (Figure 8) were obtained from Figures 6 and 7. In the blood containing tissue, hemoglobin absorption dominated the tissue difference spectra and at wavelengths less than 610 nm, absorption was too high to produce accurate spectra. On the other hand, blood-free (1% Ht) difference spectrum showed a peak at 820 nm together with 720 nm and 605 nm absorption peaks (Figure 8, bottom). The best cytochrome aa_3 spectrum were obtained *in vivo* through the skull of a blood-free rat brain (Figure 9, bottom). In comparison to *in vitro* difference spectra of purified cytochrome aa_3 from beef heart Figure 9, middle), there were equal absorbency changes at 710 nm and 1000 nm in purified cytochrome aa_3, while cytochrome aa_3 measured in the cat brain had a different absorbency change at these wavelengths. An important observation was that the purified enzyme peaks at 830 nm and 715 nm shifted to 820 nm and 730 nm in the cat and rat brain study. Differences between the *in vivo* and *in vitro* spectra of cytochrome aa_3 may be attributed to a scattering factor as well as changes in residual hemoglobin, brain cells, water, and unknown molecules that altered the absorption characteristics by the oxygen tension.

Multicomponent Curve Fitting Analysis of *In Vivo* NIR Spectra

Multicomponent curve fitting analysis was applied to *in vivo* difference spectra during normoxia and hypoxia using the extinction coefficients of oxy-and deoxyhemoglobin and reduced and oxidized cytochrome aa_3. The rationale of this technique is based on the fact that although there are changes in peak characteristics of the tissue pigment absorption spectra, the difference spectra at 700-900 nm were similar. Changes due to scattering mediums may be within experimental error i.e. noise. From the point of statistical accuracy, the more data points yield less error and this method allows evaluation with statistically significant numbers. The difference spectra between normoxic and hypoxic conditions were fitted in Figure 8, upper. The fitted data shows that the difference spectra contains 1555 units of deoxyhemoglobin, while 891 units of oxyhemoglobin were lost resulting in a total increase of 664 units of hemoglobin in the tissue. Six units of oxidized cytochrome aa_3 were lost and 2 units of reduced cytochrome aa_3 were formed. A unit is defined as the concentration times the path length. Assuming that the mean light path is 5 cm we can then calculate the change of cytochrome aa_3 and hemoglobin as a loss of 178 μM of oxyhemoglobin and a 311 μM increase of deoxyhemoglobin with a 132 μM net increase in hemoglobin concentration during short hypoxia (2 minutes). The total change of cytochrome aa_3 was 1.6 μM to 2.4 μM at that time. The time course of these changes during short hypoxia showed an immediate hemoglobin increase due to the dilation of precapillary and capillary beds. A small amount of cytochrome aa_3 was reduced shortly after the appearance of deoxyhemoglobin. It is interesting to note that the total absorption of cytochrome aa_3 decreased accordingly as the light path length shortened resulting in less brain tissue in the light path.

The absolute spectra in Figures 6 and 7 show distinct peaks corresponding to oxy- and deoxyhemoglobin, oxidized and reduced cytochrome aa_3, water, and lipids. Difference spectra (raw spectra - fitted spectra) also had distinct water peaks at 980 nm, 840 nm, and 745 nm (all downward) and residual deoxyhemoglobin peaks at 650 nm and 760 nm (Figure 8). Curve fitting analysis can be improved using known extinction coefficients but this technique must take into consideration changes in scattering as well as spectra resulting from unknown protein molecules in cells and other and other molecules that may

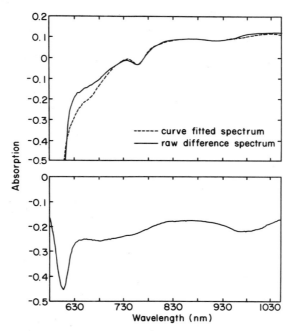

Fig. 8. Difference spectra of a cat brain during normoxic-hypoxic transition. Spectra taken at normoxia were subtracted from that during hypoxia at normal hematocrit (top figure from Figure 6) and 1% hematocrit (bottom figure from Figure 7). The dashed line is a fitted spectrum of the difference spectra. The difference between the raw and fitted curve mainly consists of water and deoxyhemoglobin peaks. A large discrepancy at lower wavelengths was observed. Two minutes of short hypoxia caused low oxyhemoglobin and high deoxyhemoglobin with a net increase of hemoglobin.

change during hypoxia. Fitting hemoglobin-free brain spectra was difficult as indicated by statistical error. The difficulties in fitting this data were the result of the low levels ofcytochrome aa_3 which were overshadowed by the large contributions of water and other molecules. It was found that water peaks changed the actual peak position from 830 nm to 825 nm, and from 715 nm to 725 nm. In addition, water peaks seem to be altered *in vivo*.

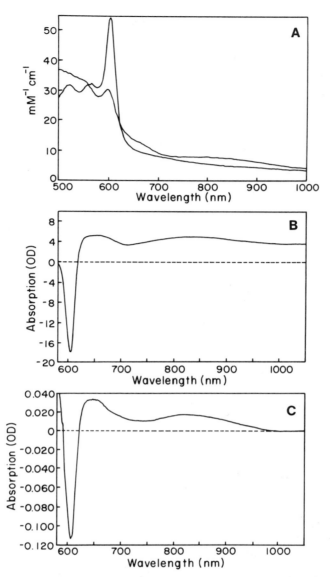

Fig. 9. A: Reduced and oxidized spectra of purified cytochrome aa_3. B: Difference spectra of top figure. C: Difference spectra of blood-free rat brain taken at oxygen fractions of 100% and 0% in the breathing gas.

REFERENCES

1. F. F. Jobsis, Non-invasive, infrared monitoring of cerebral and myocardial oxygen sufficiency and circulatory parameters, *Science*, 198:1264 1267 (1977).
2. B. Chance, J. S. Leigh, H. Miyake, D. S. Smith, S. Nioka, R. Greenfield, M. Finnander, K. Kaufman, W. Levy, M. Yong, P. Cohen, H. Yoshioka and R. Boretsky, Comparison of

time-resolved and -unresolved measurements of deoxyhemoglobin in brain,*Proc. Natl. Acad. Sci.*, 85:4971-4975 (1988).

3. D.M. Bensen and J.A. Knopp, Effect of tissue absorption and microscope optical parameters on the depth of penetration for fluorescence and reflectance measurement of tissue samples, *Photochem. Photobiol.*, 39:495-502 (1984).

4. B. Blumberg, Light propagation in human tissues: The physical origin of the inhomogeneous scattering mechanisms, *Biophys. J.*, 51:288a (1987).

5. P. van der Zee and D.T. Delpy, Simulation of the point spread function for light in tissue by a Monte Carlo method, *Adv. Exp. Med. Biol.*, 215:179-191 (1987).

6. B.C. Wilson and C.T. Adam, A Monte Carlo model for the dependence of cellular energy metabolism, *Arch. Biochem. Biophys.*, 195:485-493 (1983).

7. R. Wodick and D.W. Lübbers, Ein neues Vergahren zur Bestimmung des oxygenierungogrades von Hämoglobins-pektren bei inhomogenen Licht wegener Läutert under analyse von spektrem der menschlichen Haut,*Pflunger arch.*, 342:41-60 (1973).

8. K. Kubelka and F. Munk, Ein beitrag zur optik der frabanstrich, *Z. Tech. Phys.*, 11a:593-603 (1931).

9. D.W. Lübbers, and J. Hoffmann, Absolute reflection photometry at organ surfaces, *Adv. Physiol. Sci.*, 8:353-361 (1981).

10. C.B. Cairns, D. Fillipo and H.J. Proctor, A non-invasive method for monitoring the effects of increased intracranial pressure with near-infrared spectrophotometry, *Surg. Gynecol. Obstetric.*, 161:145-148 (1985).

11. B. Chance, Spectrophotometric observations of absorbance changes in the infrared region in suspension of mitochondrial and in submitochondrial particles, in: "Biochemistry of Copper," J. Peisach, P. Aisen and W.E. Blumberg, eds., Academic Press, New York (1966).

12. M. Cope, D.T. Delpy, E.O.R. Reynolds, S. Wray, J. Wyatt and P. van der Zee, Methods of quantitating cerebral near-infrared spectroscopy data, *Adv. Exp. Med. Biol.*, 222:183-189 (1987).

13. I. Gianni, M. Ferrari, A. Carpi and P. Fasella, Rat brain monitoring by near-infrared spectroscopy: An assessment of possible clinical significance,*Physiol. Chem. Phys.*, 14:295-305 (1982).

14. O. Hazeki and M. Tamura, Quantitative analysis of hemoglobin oxygenation state of rat brain *in situ* by near-infrared spectrophotometry, *J. Appl. Physiol.*, 64:796-802 (1988).

15. K. Kariman and D.S. Burkhart, Heme-copper relationship of cytochrome oxidase in rat brain *in situ*, *Biochem. Biophys. Res. Commun.*, 126:1022-1028 (1985).

ULTRASHORT LASER PULSES TOMOGRAPHY IN SEMI-OPAQUE MEDIA

P.P. Ho, P. Baldeck, K.S. Wong, G. Zhang, F. Raccah, G. Tang, and R.R. Alfano

Institute for Ultrafast Spectroscopy and Lasers
Photonics Application Laboratory
Department of Electrical Engineering
The City College of New York
New York, NY 10031

ABSTRACT

Spatial distribution of hidden foreign objects in a semi-opaque environment has been investigated using ultrashort laser pulse time resolved techniques with spatial resolution down to less than 1 mm. Using femtosecond streak camera ranging, 1-dimensional distance has been measured. Time sequences of 2-dimensional images have been obtained using picosecond-Kerr gate.

INTRODUCTION

Over the past several years, there has been considerable interest by the medical community in detecting hidden objects in semi-opaque media with spatial resolution of less than 1 mm. Ultrashort laser pulses have been applied to ranging[1] and 3-dimensional imaging[2-4] in random media[5-7] and biological samples[8]. Using x-ray or NMR techniques, the spatial resolution in diagnosing cancer formation in the human kidney is greater than several millimeters. In most cases, for a tumor of this size, the number of cancer cells exceeds 10^9. It is the dream of most physicians that a laser endoscope can be used for the early diagnosis of human cancers in the breast, brain, kidney, etc., in vivo. After identifying the precise location of the tumor, another pulse laser can be applied to remove the cancer tissues without performing surgery to remove the organ. Another possible application will be imaging the fluid distribution of cochlec in the inner ear. There are millions of people suffering hearing disorders due to the malfunction of cochlec. The ultrashort laser pulse imaging technique may provide a nondestructive and noninvasive method to map the inside fluid distribution of cochlec covered by a shell in humans. In the

Photon Migration in Tissues
Edited by B. Chance
Plenum Press, New York

case of material structure diagnosis, it may be used as a direct, nondestructive, in situ measurement of an object's surface contour and internal structure with submillimeter spatial accuracy.

Optical techniques can provide spatial information from signals by several different processes: diffusive scattering, specular reflection, absorption, fluorescence, and Raman scattering, to name a few. In the first three processes, the signal wavelength remains the same as the incident pulse wavelength, while in the latter two cases, new wavelengths are emitted from the excitation process.

There are several approaches to apply laser technology to readout a 3-dimensional object. One can set the object to be stationary and use a scanning laser beam to probe the entire space. Secondly, one can set the laser beam at a fixed direction and move the object. Thirdly, if new wavelengths are generated by the sample, the incident laser beam can be expanded to cover the entire space to be investigated. The laser beam requires no moving parts.

In this presentation, direct time resolved measurements of the 1-dimensional femtosecond laser pulse ranging and the 3-dimensional picosecond Kerr imaging in semi-opaque water solutions will be discussed. This work is the forerunner of actual optical imaging in the human body. In both cases, signals were obtained from the scattering process.

FS-PULSE RANGING

Experimental Methods

The ranging experiment setup is shown in Figure 1. It consists of a femtosecond-laser system, a water cell, a streak camera system, and a fiber probe. The spatial information from the temporal measurement is obtained by using the conversion $\Delta x = c \, \Delta t$. The time resolution of the light scattering measurements is affected by the incident laser duration, streak camera resolution, fiber collection spatial window, group velocity dispersion of light in optical fiber and water, incident laser beam spot size, and multiple scattering inside the random medium. Details of these items will be discussed as follows:

Laser: Laser pulses were generated from a CPM dye laser[9] with an output pulse duration of 200 fs, 20 mW average power, and 625 nm wavelength at 114 MHz repetition rate. In the single shot operation, a single pulse selected out of 114 MHz laser pulses was amplified by a high power YAG laser to achieve a 1 mJ per pulse at 20 Hz repetition rate[7]. In the latter method, the time resolution was improved. However, the S/N was greatly reduced due to the significant decrease in the number of signal pulses. These pulses were collimated to a beam spot size of < 1 mm over the entire measurement path of the water cell.

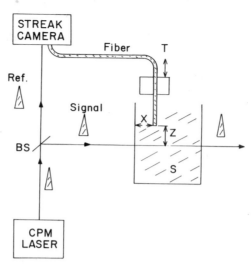

Fig. 1.   Experimental apparatus of fs-pulse-fiber-streak camera ranging.   Typical experimental
error was ± 2 ps. CPM: colliding-pulse-mode locked dye laser system with YAG laser
amplifier; BS: beam splitter; S: sample water cell with polystyrene particles; SC: streak
camera; X: horizontal distance of laser pulse travels from the window of the cell to the
fiber collector; Z: vertical distance from the laser beam to the end surface of the fiber
collector; T: translational stage to control the vertical location of fiber.

Semi-opaque media: An optical cell was constructed using a 1/4" thick plexi-glass
plate of a dimension ~6"x4"x4". The large dimension of the outside boundary reduced the
reflection and scattering from the water cell surfaces.   Polystyrene particles were
suspended in water with volume fraction 0.1 with average diameter of 2.95 ± 0.16 μm. Dry
milk powders were used to simulate the semi-opaque environment during the pre-alignment
process.

The optical density coefficient, ODC (cm^{-1}) = OD/optical path = $\alpha/2.3$ which
includes the loss from both scattering and absorption[10] at 625 nm wavelength was
measured through a 1 cm long cell by a Perkin-Elmer Lamda[9] spectrophotometer. For the
polystyrene particle water solution, the absorption coefficient is negligible in comparison
to the scattering coefficient. By diluting the purchased polystyrene particles by 10^{+4} times
which yielded a volume fraction of f+10^{-5} in water, an ODC ~0.5 cm^{-1} was obtained. At this
ODC = 0.5 cm^{-1}, there were ~4 x 10^5, 2.95 μm polystyrene particles in 1 cm^3 water
solution. Another factor which has been noticed is the wavelength dependent on the opacity.
At longer λ of 750 nm, the OD is about 30% higher than that at 625 nm. The lowest value of
OD in the visible spectrum is ~600 nm. The OD increased again at a shorter wavelength.
The OD at 450 nm is about 20% higher than the OD at 600 nm.

Streak Camera Technology: Multi-mode optical fibers without lenses were used to
collect the vertical scattered laser pulse and transmit the signal to a Hamamatsu model

C1587 streak camera system with either a single shot #1952 module or a sync-scan (M1955 and M2567) module and blanking unit. Temporal information of laser pulse propagation and scattering in time was measured by the streak camera and digitized, stored, and processed by an IBM-PC-XT computer.

A reference pulse which travels through a fixed distance in air is also directed into the streak camera to set the starting time of the streak camera and also to display the minimum system time-resolution. By measuring the time separation between the peak of the reference pulse to the signal pulse, the pulse propagation time or distance information can be measured. After deconvoluting the temporal broadening factors, the width measurement of the signal pulse can extract the dimension of the object.

The high sensitivity of the sync-scan streak operation was a key factor in determining the dynamics of photon migration. About 100 million scattered laser pulses were collected each second in the sync-scan streak camera operation. In a highly opaque medium, the total accumulated streak signal may be only a few thousand counts (1 count ~10 photons). This is equivalent to about 10^{-3} photons per pulse being scattered and detected by the streak camera input. The sync-scan streak camera method is an extraordinarily sensitive time resolved technique to detect scattered light in semi-opaque random material and biological systems.

The optimum time resolution is ~8 ps for sync-scan mode and ~2 ps for single shot operation of the streak camera. In the sync-scan operation, the S/N ratio was greatly improved due to the significant increase in the number of signal pulses to be accumulated.

Fiber Collection: A multi-mode optical fiber with an inner core diameter of 100 μm was immersed in the water solution to collect the scattered laser pulses. The signal collection and transmission fiber received all light scattered within the collection cone angle specified by the numerical aperture (NA). The NA value was specified by the fiber manufacturer. Fibers with NA = 0.3 (air) were purchased from Newport Corporation and those with NA = 0.2 (air) were obtained from Spectran Corporation. In our experimental condition of water immersion, the effective NA in water is NA (air)/n, where n is the index of refraction of water. At 625 nm, both the group and phase refractive index of water are approximately the same at ~1.33.

The geometric property of a fiber collection introduces an added time broadening factor for a scattering measurement scheme as shown in Figure 2. The vertical distance of the collection fiber and surface tip to the laser beam is as shown. The overall time bandwidth due to the total collected scattering signal from the entire path length AB is $\Delta T = AB/(c/n) \sim 2MA(z/c)$. An optical fiber with NA = 0.2 situated at z = 10 mm from the laser beam will introduce a time band of $2 \times 0.2 \times 1/3 \times 10^{10} = 13.3$ ps. This time resolution was verified by our measurements.

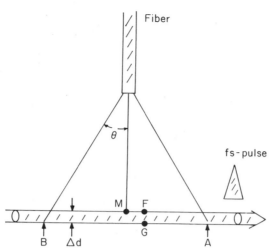

Fig. 2. A schematic of the scattering signal collection using an optical fiber. θ: collection cone angle = $\sin^{-1}(NA/n)$; NA: numerical aperture of fiber in air; n: index of refraction of the medium; Δd: laser beam diameter; ΔT (FWHM of the scattered pulse) = $AB/(c/n)$ = $2z$ $\tan\theta/(c/n)$ ~ $2zNA(1/n)/(c/n)$ = $2z\,NA/c$. M: center point of the collection cone.

Dispersion: Group velocity dispersion and geometrical dispersion are two common factors for the pulse broadening in multi-mode fibers. The material dispersion coefficient[9] for a high quality optical fiber at 625 nm wavelength is ~0.3 ps/nm-m. For example, a transform limited 100 fs pulse with ~4 nm spectral banwidth traveling through a 1 m long optical path in fiber or water will broaden to ~3 ps over the other hand. The geometrical dispersion in a 100 μm core fiber with NA = 0.2 is about 44 ps/m. In our measurement, the total length of the fiber from the collection tip to the input slit of the streak camera was set to be about 20 cm. The time broadening effect was about the minimum resolution of the sync-streak mode.

Beam Spot Size: An additional temporal broadening process occurs from the beam spot size in our vertical geometric arrangement. As shown in Figure 2, the upper part of the laser beam at point F will take $\Delta d\, c/n$ less time to reach the fiber tip than the lower part from the bottom point G. A beam size of $\Delta d=1$ mm introduces a 4.4 ps time broadening factor.

Multiple Scattering: As the random medium becomes extremely dense, multiple scattering occurs which causes photons to be scattered back and forth inside the water solution before reaching the fiber. Thereby, the total optical path is increased which broadens the temporal profile of the signal pulse[3,4].

The incident laser pulse duration, streak speed, fiber length, and beam spot size were kept the same in measurements in section II. The time profile of the signal pulse occurred by changing the fiber collection window size (vertical distance) and the number of scattered particles.

RESULTS

The time dynamics of the fs-pulse migration in a random medium collected by a fiber and dispersed in the single shot and the sync-scan modes are displayed in Figures 3 and 4, respectively. In both figures, the leading pulse is the reference pulse. By measuring the difference in the peak arrival time between the reference pulse and signal pulse, light migration process in time has been analyzed over different vertical distance and number of scatters.

Under single shot operation of Figure 3, the time resolution of the entire system was about 2 ps. Many random spikes arose from electronic noise or from the scattering process of the random medium. Under the sync-scan operation of Figure 4, the time resolution was ~8 ps. The signal profile was much smoother due to the accumulation of over 100 million pulses. The temporal profile of the vertical scattered signal as a function of fiber location is displayed in Figure 4 at a fixed ODC ~1 cm^{-1}. At z = 0 cm, the scattered signal is symmetrical (Figure 4a). The shape is almost the same width as the reference pulse. When z increased, the peak signal was delayed and an asymmetrical multiple scattering tail was observed as shown in Figures 4b to 4d.

A metal pin acts as a phantom in the semi-opaque environment. The scattered signal from a metal pin is displayed as a function of the opacity of the water solution in Figure 5.

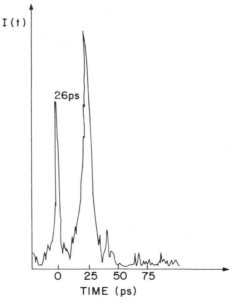

Fig. 3.   Temporal profile of vertical scattered fs-laser pulse measured by a single shot streak camera at z = 4 mm and ODC ~ 0.5 cm-1. The vertical scale is the intensity distribution of the scattered signal in arbitrary unit. The horizontal axis is the time in ps. The leading (left) pulse is the reference pulse propagated in a fixed air distance to set up an absolute time scale.

In the clear water (OD-0) of Figure 5a, the pin is easily identified with an excellent S/N value. When opacity of water solution was increased by adding polystyrene particles, the background particles scattering increased and S/N decreased. Up to ODC = 1.64 cm^{-1} (total OD ~3.28 for 2 cm path which includes ~1 cm incident light path and 1 cm scattering path), the S/N was down to ~1.5 as shown in Figure 5c. The small spike located at near the right center of the signal pulse in Figure 5c is the pin. This can be verified in Figure 5d where the spike disappeared when the pin was removed.

One notices that scattered signal/reference pulse also decreased as OD increased. In this sequence of measurement, the sensitivity of the streak camera was increased by a factor of ~100 times to compensate the high loss of a higher OD. Accordingly, the reference pulse was attenuated to offset the electronic gain. The signal occurs before the pin peak (Figures 5b and 5d) arises from the background scattering of the random medium.

The scattered signal from two metal pins separated by 3.5 ± 0.5 mm in the water solution is displayed in Figure 6. The two peaks were separated by ~16 ps which is in

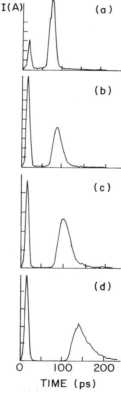

Fig. 4.   Temporal profiles of vertical scattered fs-laser pulses measured by a sync-scan streak camera at various fiber locations: z: (a) 0 mm, (b) 5 mm, (c) 10 mm, (d) 20 mm. T(peak): time separation between the peak of reference pulse to the signal pulse; ΔT: FWHM of the signal pulse.

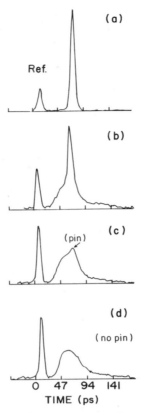

Fig. 5. Optical pulse ranging of a metal pin in water solutions of different opacity at a fixed z = 10 mm. (a) OD = O (clear water), (b) OD = 0.87 cm^{-1}, (c) OD = 1.64 cm^{-1}, (d) OD = 1.64 cm^{-1} (metal pin removed).

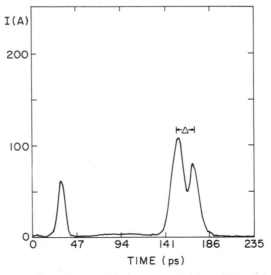

Fig. 6. Optical pulse ranging of two metal pins separated by ~ 3.5 mm in water. Δ ~ 16 ps.

excellent agreement with the calculated ΔT = 3.5 mm/c/n ~15.5 ps. This double pin scattering measurement shows the feasibility of mm-spatial resolution optical ranging using the fs laser pulse and streak camera method.

ANALYSIS

In a semi-opaque water solution of ODC ~1 cm^{-1}, the arrival time and the FWHM of the scattered signal pulse are plotted as a function of the vertical distance from the collection fiber tip to the laser beam in Figures 7 and 8, respectively. In Figure 7, the peak arrival time of the collected signal was found to be linearly dependent on the vertical separation distance from the fiber to the laser beam. Data can be fitted with a straight line with a calculated slope = $\Delta t / \Delta z$ = n/c = 4.12 ps/mm. The linear regression factor R is ~1. The calculated slope in Figure 7 is most unusual. The slope should be the inverse of group velocity of the laser pulse. The calculated n from our measured data is 1.24 which is significantly smaller than the refractive index of water = 1.33. Both measurements and calculation errors are smaller than this discrepancy.

The diffusion constant, D, is proportional to the mean free path, ℓ. With ℓ = 1/Nσ = 1/OD where N is the number density of polystyrene particle, and σ is the total scattering cross section of a particle. When N is increased, the opacity of water OD increases, and and D decrease. The peak of scattered pulse shown in Figures 5 and 10 follows this dependence. At lower particle concentration when the mean free path, ℓ, is long, the vertical scattered photons travelling shorter z distance will not experience any additional scattering. This looks like a ballistic process for the scattered photons collected by the fiber. When > 20-mm, the FWHM of the scattered pulse in ballistic process has little influence on the multiple-scattering (diffusion) process. The broadening of the pulse width mainly originates from the fiber collection NA.

In the vertical scattering collection geometry, due to the scattering loss, the incident number of photons decreases in the forward propagation direction x. The measured I(t) within a finite collection cone angle should be written as I(t) = I_o(t) exp (-αx), where I_o(t) is a symmetrical scattering function by assuming a constant incident photon flux where the total extinction coefficient α -> 0. The peak of the I_o(t) signal will mainly arrive from the center of the collection cone or the M-point in Figure 2.

When α > 0, then the peak of I(t) will shift toward the left and arise from the point at the left side of point-M. This shift towards the early time indicates a quasi-ballistic (single scattering) event. The total optical path along the x-direction as well as the travelling time have been reduced with the increasing of scattering particle concentration. The faster arrival of scattered signal originates from the reduction of propagation optical path. This is "not" a decreasing of the index of refraction as it apparently looks in Figure 7. In the case of very dense media where OD is very high, multiple scattering or diffusion process dominates. Under this condition, the T (peak) profile can be fitted by a quadratic function.

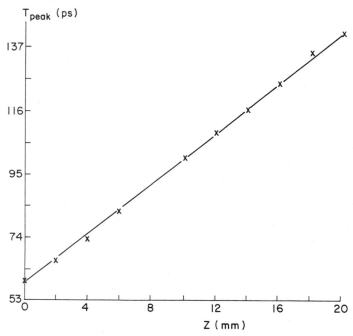

Fig. 7. Measured peak arrival time of scattered fs-pulse in a semi-opaque water OD = 0.5 mm as a function of fiber location. T_D: normalized delay time of peak scattered signal for $T_D = 0$ at z = 0. The solid line is a linear least square fitted straight line with slope = 4.12 ps/mm.

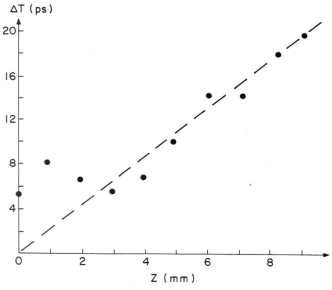

Fig. 8. Measured scattered pulse width in a semi-opaque medium as a function of z. The solid line is a calculated curve of $\Delta T = 2z\ NA/(c/n)$.

In Figure 8, the FWHM of the signal pulse shows a small plateau and bump in the region $z = 0$ to 3 mm. This bump most likely arises from the system resolution and the possible multiple scattering between the fiber end surface and the random particles. As $z > 3$ mm, the FWHM of the scattering pulse increased monotically. These data can be fitted by a straight line with slope ~1.8. This slope is predicted within our experimental error of ~10% by the theoretical fiber window broadening equation $\Delta T/z$ ~2 NA/(c/n) = 1.78 when NA = 0.2 and n = 1.33.

3-D KERR IMAGING

Experimental Methods

An optical Kerr gate[2,3] pumped by ultrashort laser pulses can be used as an ultrafast optical camera shutter. When an object is illuminated by an ultrashort laser pulse, there will be a scattering signal of the object towards the shutter and the 2-dimensional detection system. Since the time for light travel in space takes $t = x/c$, the scattered signal from each different segmental plane of the object will arrive at the detection plane at different times. The signal from the front part of the object will reach the detector earlier than those from the rear part of the object. That is, we can optically slice a 3-dimensional object into multiple 2-dimensional segmental plane blocks. The third spatial coordinate corresponds to the time coordinate.

When the pump pulse is synchronized in time with a particular plane of the object, only this particular segment can be imaged at the detector plane. The rest of the object will be removed by the Kerr shutter. In this manner, if there is a foreign particle hidden inside the object, the outside opaque shielding can be removed. Therefore, direct observation of the inside structure can be imaged. Since the removal of noise of the Kerr gate is based on the extinction ratio of a pair of crossed polarizers, a typical Kerr shutter can obtain the S/N ratio of about 10^4, which means that the total scattering noise from the background environment or other sources cannot be greater than 10^4. In addition, noise generated from the optical path of the scattered signal is intrinsically difficult to remove from the time-gate method.

The spatial resolution of a Kerr gate depends on the pump pulse duration, response time of the Kerr medium, and overlapping angle and convolution function between pump and probe pulses. Using a CS_2 Kerr medium with molecular relaxation time of 2 ps and pump and probe laser pulses of 10 ps duration, the overall system resolution (FWHM) is about 17 ps[3].

The experimental arrangement for an ultrafast Kerr shutter is shown in Figure 9. It consists of a mode-locked glass laser system which emits 1.06 μm and 0.53 μm pulses of 8 ps duration, a Kerr shutter, and a 2-dimensional readout system. The basic principle behind the time shutter was demonstrated by Duguay[2] in 1970 for the time of flight of a ps laser pulse in milky solution. The major differences between our system and Duguay's 1970 Kerr shutter is the readout system. A 2-dimensional CCD camera with 600 x 400 spatial resolution is used to record 2-dimensional images. An IBM-AT compatible computer with 1MB RAM and 30MB hard disk is used for signal processing and data storage. A 12-bits resolution frame grabber by Data Translation, Inc. is used to digitize the 2-dimensional image, and the data is processed with an imaging analysis software package.

RESULTS

A set of time sequence (spatially segmented) pictures of a metal screw situated in air is displayed in Figure 10. A clear image of the screw is displayed in Figure 10a when the gating pulse and the screw scattered signal pulse are synchronized. Delaying the arrival

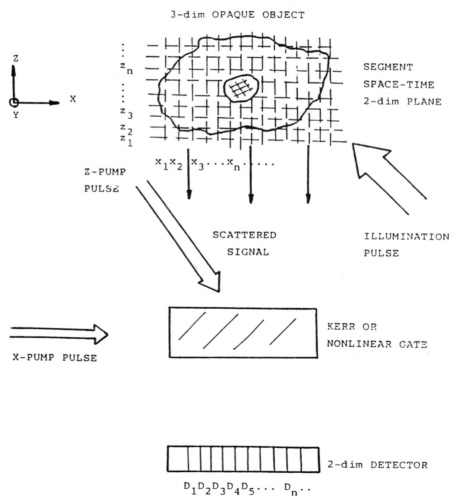

Fig. 9. The experimental setup of a 3-dimensional Kerr imaging system.

time of the gate pulse by 10 ps, 20 ps, and 40 ps, the image of the screw gradually fades away as displayed in Figures 10b, 10c, and 10d, respectively. In Figure 10d, the Kerr shutter blocks entirely the total detectable screw scattered signal.

Another set of Kerr images of a screw located in a milky water solution is displayed in Figure 11. A sequence of images for ungated (cross-polarizers have been turned to be parallel) images of a screw, a milky water cell, and a milky water cell and a screw are displayed in Figures 11a, 11b, and 11c. In Figure 11c, the image of a screw is blurred by the surrounding milky particles. With a Kerr shutter, the image of a screw in milky water is clearly identified as shown in Figure 11d. Most of the background light scattering has been shielded out.

SUMMARY

One-dimensional pulse ranging in random media has been demonstrated using fs-laser pulse, fiber collection, and sync-scan streak camera. Quantitative analysis of photon

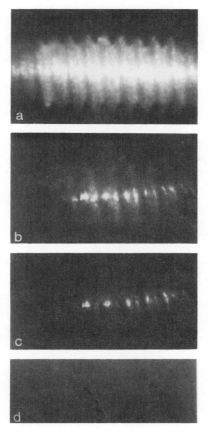

Fig. 10. Kerr images of a screw situated in air. The time difference of the arrival time between the gated pulse and the screw scattered signal: $\Delta\tau =$ (a) 0 ps, (b) 10 ps, (c) 20 ps, (d) 40 ps.

dynamics in vertical scattering geometry have been performed as a function of distance and opacity providing direct information on how the laser pulse propagates and scatters in semi-opaque random media. Feasibility of quantitative data on mm-ranging has been successfully demonstrated. The 3-dimensional Kerr image technique using a modern readout image system has been performed. We can digitize a segmental 2-dimensional plane from a 3-dimensional object and then possibly reconstruct using data obtained form different time sequences. Hidden objects in semi-opaque media of total optical density less than 3 can be identified. Using multiple wavelength for probing and using the possible fluorescence and absorption detection schemes, the signal to noise ratio can be greatly improved for possible imaging in vivo. More basic studies of light propagation and scattering are needed in order to develop optical tomography of organs in the human body. These measurements and analysis will have an impact on the understanding and practical applications for the use of light in optical medical imaging technology.

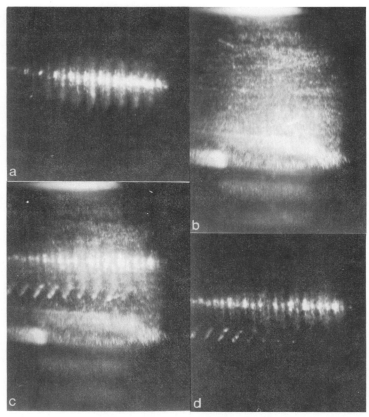

Fig. 11.   Comparison of images of a metal screw in milky water with and without Kerr gates.   (a) Screw in air without gate, (b) Milky water cell without gate, (c) Scew in milky water without gate, (d) Screw in milky water with gate.

ACKNOWLEDGEMENTS

This research is supported in part by SDIO, NYSST, PSC/CUNY, NIH, and Hamamatsu Photonics K.K.

REFERENCES

1. J.G. Fujimoto, S. De Silversti, E.P. Ippen, R. Margolis, and A. Oseroff, Femtosecond optical ranging in biological systems, *Opt. Lett.*, 11:150-2 (1986).
2. M.A. Duguay and A.T. Mattick, Ultra high speed photograph of picosecond light pulses and echoes, *Appl. Opt.*, 10:2162-70 (1970).
3. P.P. Ho and R.R. Alfano, Optical ken effect in liquids, *Phys. Rev.*, A20:2170-82 (1978).
4. K.G. Spears, T.J. Robinson and R.M. Roth, Particle distributions and laser particle interactions in an RF discharge of silane, *IEEE, Trans. Plasma Sc.*, PS14:179-87 (1986).
5. G.H. Watson, P.A. Fleury and S.L. McCall, Search for photon localization in time domain, *Phys. Rev. Lett.*, 58:945-7 (1987).
6. G. Yoon, A.J. Welch, M. Motamedi and M. Van Gemert, Development and application of three-dimensional light distribution model for laser irradiated tissue, *IEEE.*, J-QE23:1721-33 (1987).
7. W. Jiang, D. Sun and F. Li, Particle behavior of femtosecond optical pulses, *Opt. Comm.*, 66:152-4 (1988).
8. B. Chance, Photon Migration Workshop, University of Pennsylvania, April 17, 1988.
9. P.P. Ho, A. Katz, R. Alfano and N. Schiller, Time resolution of ultrafast streak camera system using femtosecond laser pulses, *Opt. Comm.*, 54:57-62 (1985).
10. H. van de Hulst, Light scattering by small particles, John Wiley & Sons, New York (1957).
11. D. Gloge, E.A. Marcatili, D. Marcuse and S.D. Personick, Dispersion properties of fibers, in: "Optical Fiber Telecommunications," S. Miller and A. Chynoweth, eds., Academic Press, New York (1979).

BREAST BIOPSY ANALYSIS BY SPECTROSCOPIC IMAGING

Clyde H. Barlow[a], David H. Burns[b], and James B. Callis[c]

[a]The Evergreen State College
Olympia, Washington 98505 and
Department of Chemistry, University of Washington
Seattle, Washington 98195

[b]Center for Bioengineering, University of Washington
Seattle, Washington 98195

[c]Department of Chemistry, University of Washington
Seattle, Washington 98195

INTRODUCTION

Short wavelength near infrared (SW-NIR) spectroscopy possesses many features that are advantageous for spectroscopic studies of tissue anatomy and physiology. Human tissue is relatively transparent in the 650-1300 nm spectral region so that relatively thick samples can be examined. At the same time one observes spectral features which arise from low lying electronic transitions of heme proteins and vibrational overtones from major constituents of tissue. As an example, reflectance oximetry[1] is based upon low energy electronic transitions of oxy and deoxyhemoglobin, Figure 1. The spectrum of the hemoglobin molecule is very different in the two forms. In oximetry, hemoglobin saturation is determined non-invasively by measuring the light scattered from tissue at 650 and 805 nm; the former wavelength represents a maximum in the absorbance difference between the oxy and deoxy forms while the latter represents an isosbestic point for the two forms. A substantial advance in oximetry was made by Yosohiya[2]. In his scheme, recognition was taken of the fact that the light reflected from or transmitted through human tissue exhibits both a steady state (DC) and fluctuating (AC) signal. The former arises from tissue pigments and venous blood, while the latter arises from arteriole blood and fluctuates synchronously with the heart rate. Jobsis advanced the oximetry concept one step further. He noted that in

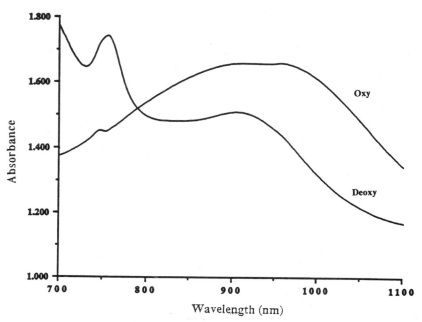

Fig. 1. SW-NIR spectrum of oxygenated and deoxygenated packed red blood cells versus air at 1.6 mm pathlength. Principal absorbance bands are due to low energy electronic transitions in oxy- and deoxyhemoglobin. Spectra were recorded on a Pacific Scientific Model 6250 NIR spectrophotometer.

addition to hemoglobin, copper in cytochrome oxidase, the terminal enzyme of the mitochondrial electron transport chain, could also be measured in tissue by SW-NIR spectroscopy. This work demonstrated the possibility that oxygen availability in the brain can be measured non-invasively by transillumination spectroscopy in the near infrared[3]. A major conceptual breakthrough in oximetry has recently been demonstrated by Chance[4]. Using time resolved pico-second spectroscopy Chance has measured localized hemoglobin dependance in tissues and has laid the foundation for NIR tomography of hemoglobin concentration and saturation.

Another source for absorption bands in the SW-NIR arises from overtones of the fundamental vibrations which originate in the mid infrared. For natural products, the observed peaks are largely due to overtone stretches of OH, NH, and CH groups; other groups may be represented by combination bands. In the SW-NIR, the absorption bands of each chemical grouping are observed a number of times as the second through third or fourth overtones, but owing to the selection rules, the bands become progressively weaker at shorter wavelengths. As a consequence, the total peak density tends to be higher than in the mid-IR, giving more spectral overlap. At the same time, the absorption bands in the SW-NIR are at most 1% and typically only 0.1% of the intensity of their mid-IR counterparts.

The earliest studies of the analytical utility of the vibrational NIR were reported by W. Kaye[5,6]. However, given the disadvantages of multiple overlapping peaks and weak absorption strengths, interest in NIR declined markedly in the 60's and 70's[7].

Nevertheless, Karl H. Norris and coworkers at the U.S. Department of Agriculture successfully demonstrated the potential of NIR reflectance spectroscopy combined with multivariate statistical analysis for characterization of natural products of agricultural importance[8]. This potential, further enhanced by the availability of extremely sensitive detectors can produce high signal-to-noise ratio spectra (20,000) eliminating the need for expensive, delicate Fourier Transform techniques. Indeed signal-to-noise is so great as to allow spectroscopic imaging with multidimensional detector arrays. Clearly, NIR spectroscopy is a technique which combines the functional group sensitivity of infrared with the more permissive sampling characteristics of the visible wavelength region.

Multispectral image analysis has previously been utilized in our laboratories to quantitatively analyze thin layer chromatography plates with overlapping absorption bands[9,10]. In this paper we present progress on SW-NIR multispectral image analysis of a breast tissue biopsy.

INSTRUMENTAL DESIGN

Multispectral Scanner

Our experimental apparatus was similar to the previously described multichannel imaging spectrophotometer developed for thin layer chromatography[9,10]. An intensity regulated 50-W tungsten filament lamp, served as a light source as shown in Figure 2. An image of the filament was projected onto the entrance slit of a 0.32-m monochromator (Instruments SA H-320) equipped with an 1200 line/mm grating blazed for 1000 nm. The slit width of the monochromator was set to provide a 5 nm spectral band-pass. A 250 mm focal length achromatic doublet was used to project an image of the grating onto the tissue specimen. Size of the grating image was 45 mm square. Wavelength selection was accomplished by a stepper drive motor connected to the sine bar drive of the monochromator. Light transmitted through the tissue was collected by a Kowa f/1.3 cine lens and focused onto the photosensitive surface of a SW-NIR-to-visible viewer (Electrophysics 7310). The visible light output from the viewer was imaged onto a silicon target vidicon camera (RCA TC-2000) using a pair of Cosmicar f/1.4 cine camera lenses. While the silicon target vidicon camera had sufficient sensitivity in this region to carry out the imaging studies directly, coatings on the tube face produced irregular wavelength dependent interference patterns that confounded multiwavelength analysis. The photo-sensitive surface of the image converter did not have interference problems. Output of

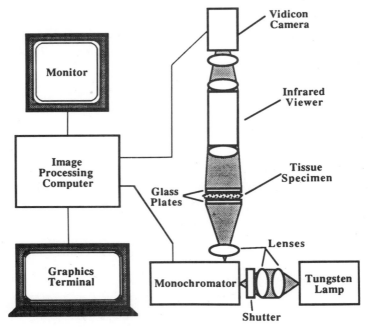

Fig. 2. Instrumental layout of near infrared spectroscopic imaging system.

the vidicon camera was digitized at video rates on a PDP-11/04 equipped with a custom built digitizer. Details of the digital television photometry system have been previously described[11].

Data Acquisition Software

To collect spectra from the SW-NIR imager, a software package was developed that allowed the acquisition and storage of tissue images while scanning the wavelength drive of the monochromator. Restriction of storage memory dictated the use of the 64 x 64 pixel mode of the buffer memory. In all of the data presented the number of frames summed at each wavelength was set to 100. When an equal number of dark frames were subtracted, this resulted in a total acquisition time for each image of approximately 3 seconds. Images were stored sequentially and indexed according to the order acquired in the spectroscopic run, allowing the individual images to be readily displayed for postrun viewing. Due to the lack of hardware floating point operations in the PDP-11/04, all postrun processing was done on a Vax 11/780. Images were displayed on a Methius Omega-420 graphics terminal and photographed with a Dunn 632 Color Camera system.

Tissue Specimen

The breast biopsy was the kind gift of Dr. Ransey Cole, New Haven General Hospital, New Haven, CT . A 4 x 4 cm breast biopsy, 5 mm thick, was stored under refrigeration and warmed to room temperature prior to analysis. The sample was mounted between parallel glass plates to ensure uniform thickness during transillumination. Two thin layer plates (Whatman LHP-K) served as a scattering transmission reference for biopsy images. Dark corrected reference and breast biopsy images were collected and stored at 5 nm steps from 650 - 1010 nm. Absorbance images were generated at each wavelength by ratioing each biopsy image with its wavelength partner reference, followed by negative log conversion. The 1.8 x 1.8 cm area of the biopsy that was analyzed filled the central 32 x 32 pixels of the video image.

RESULTS

Short wavelength near infrared contains absorbance contributions from low energy electronic transitions and overtones of fundamental molecular vibrations. Figure 1 illustrates contributions to the spectrum from oxy and deoxyhemoglobin, significant components of tissue. Compounds without low level electronic transitions also contribute to the SW-NIR spectrum. Spectra of substances which represent major vibrational contributors to the SW-NIR region are shown at the top and bottom of Figure 3. Water, the major constituent of tissue, exhibits a broad absorbance centered at 970 nm, arising from a combination of the symmetric and antisymmetric OH vibrations. Distinctly different from water, is the SW-NIR spectrum of olive oil, a representative biological glycerol fatty acid triester. Two absorption bands are observed at 930 and 1040 nm. These bands are due to overtones of methylene CH vibrations of the fatty acid side chains. Spectra obtained from various tissue composition are shown between the water and oil spectra in Figure 3. As is evident, the tissue spectra are, for the most part, a combination of the water and oil spectra. Dominating the spectra of the fat, muscle and human finger is the water band. As the relative concentration of CH increases in the various specimens, contributions from CH vibration bands also are seen to increase. The muscle spectrum shows only a small contribution from the CH vibrations and would appear to be mostly composed of water. Fat, on the other hand, has a noticeably higher CH vibration contribution with less contribution from water OH. The human finger spectrum has contributions from both CH and OH along with the broader features from oxy and deoxy-hemoglobin.

Single wavelength absorbance images are informative but are ambiguous due to overlap of spectral features. Figure 4A shows an image of breast tissue taken at 650 nm. Spectra obtained from regions indicated by arrows are shown in Figure 4B. Regional differences in SW-NIR spectra are observed. The spectra reflect tissue composition.

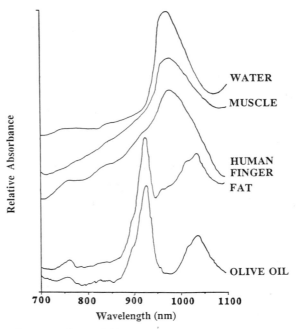

Fig. 3. SW-NIR absorbance spectra of biological samples. Spectra have been offset vertically. Overtones of fundamental C-H stretch from fatty acid methylenes at 930 and 1040 nm and overtones of fundamental O-H stretch from water at 970 nm dominate the region. Spectra were recorded on the multispectral imaging system modified for nonimaging photomultiplier detection.

Difference images were constructed to examine distribution of tissue composition by subtraction of absorbance images at wavelengths shown in Figure 5A. Wavelengths were chosen which reflected hemoglobin, OH, and CH concentrations. Only deoxy-hemoglobin was observed spectrophotometrically. This is consistent with the anaerobic condition of unperfused tissue mounted between glass plates. Deoxyhemoglobin dominates the spectrum of the tissue at 650 nm but shows little absorption at 780 nm. Figure 5B shows the difference image obtained using these wavelengths to emphasize hemoglobin distribution in the biopsy. To emphasize the water content of the tissue, the absorbance images at 780 nm was subtracted from that at 965 nm, the major OH absorbance band, and is shown in Figure 5C. Fat in the tissue was emphasized by difference of the absorbance images at 995 nm and 965 nm as shown in Figure 5D. Fat, appears emphasized as bright areas in the dark tissue background of the breast biopsy. The three images show different distributions for hemoglobin, fat, and water. In the breast biopsy sample, water appeared to have a rather

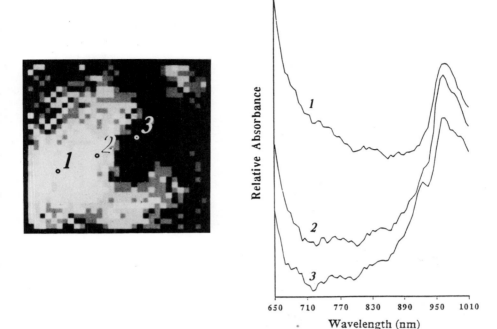

Fig. 4. (A) Single wavelength absorbance image taken at 650 nm. This image is representative of 73 images that were collected at 5 nm intervals, 650-1010 nm. (B) Absorbance spectra extracted from pixels indicated in A. Regional composition of tissue is reflected by differences in light absorption in the hemoglobin, fatty acid and water regions of the spectra.

uniform distribution, while the distribution of fat and hemoglobin show qualitatively inverse patterns. Results of the difference images are consistent with visual assessment in the biopsy.

CONCLUSION

We have shown utility of short wavelength near infrared to image tissue and emphasize composition. Images taken at wavelength corresponding to hemoglobin, OH and CH can be combined to enhance visibility of these tissue constituents. As shown by our group and others (8,9,10), application of state-of-the-art statistical analysis techniques can correlate other features important to tissue composition. These techniques may be combined

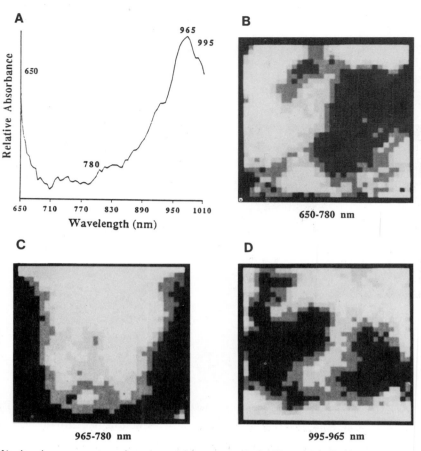

Fig. 5. (A) absorbance spectrum from breast biopsy sample in Figure 4 indicating wavelengths used to generate absorbance difference images. (B) absorbance image at 650 nm minus absorbance image at 780 nm illustrating hemoglobin distribution. (C) absorbance image at 965 nm minus absorbance image at 780 nm illustrating water distribution. (D) absorbance image at 995 nm minus absorbance image at 965 nm illustrating fat distribution. Higher intensity indicates greater concentration of constituent.

to determine tissue composition, noninvasively, and provide a valuable modality for diagnosis and screening. With the emerging technology of NIR tomography, quantitation deep in tissue of constituents other than hemoglobin may be realized.

ACKNOWLEDGMENTS

Authors are grateful to Dr. Jeffery J. Kelly for valuable discussion and assistance in manuscript preparation. Work supported by grants from Spectrascan Corporation.

REFERENCES

1. M.L. Polanyi and R.M. Heher, New reflection oxymeter, *Rev. Sci. Instrum.*, 31:401 (1960).

2. I. Yoshiya, Y. Shimada and K. Tanaka, Spectrophotometric monitoring of arterial oxygen saturation in the fingertip, *Med. and Biol. Eng. and Comput.*, 18:27 (1980).

3. F.F. Jobsis, Noninvasive, infrared monitoring of cerebral and myocardial oxygen sufficiency and circulatory parameters, *Science*, 198:1264 (1977).

4. B. Chance et al (this volume).

5. W. Kaye, Near infrared spectroscopy: A review, I. Spectral identification and analytical applications, *Spectrochem. Acta.*, 6:257 (1954).

6. W. Kaye, Near infrared spectroscopy: A review, II. Instrumentation and techniques, *Spectrochim. Acta.*, 1:181 (1955).

7. K.B. Whetsel, Near infrared spectrophotometry, *Appl. Spectros. Rev.*, 2:1 (1968).

8. K. Norris, Instrumental techniques for measuring quality of agricultural crops, in: "Post Harvest Physiology and Crop Preservation," M. Lieberman, ed., Plenum Press, New York (1983).

9. M.L. Gianelli, D.H. Burns, J.B. Callis, G.D. Christian and N.H. Andersen, Multichannel imaging spectrophotometer for direct analysis of mixtures on thin-layer chromatographic plates, *Anal. Chem.*, 55:1858 (1983).

10. D.H. Burns, J.B. Callis, and G.D. Christian, Robust method for quantitative analysis of two-dimensional (chromatographic/spectral) data sets, *Anal. Chem.*, 58:1415 (1986).

11. D.W. Johnson, J.A. Gladden, J.B. Callis and G.D. Christian, Video fluorometer, *Rev. Sci. Instrum.*, 50:118 (1979).

PHOTON MIGRATION IN MUSCLE AND BRAIN

B. Chance, D.S. Smith, S. Nioka, H. Miyake, G. Holtom and M. Maris

Department of Biochemistry and Biophysics
University of Pennsylvania
Philadelphia, PA 19104 USA

INTRODUCTION

The initial studies of highly scattering biological materials by optical means were made by David Keilin and E.F. Hartree[1,2]. Using the microspectroscope they observed that the cytochrome absorption bands from various microorganisms, which had been frozen in water or glycerol, were greatly enhanced by micro-recrystallization. Their samples were rapidly frozen in liquid nitrogen, and then rewarmed to a temperature at which a "red halo" was visible in the transmitted light. This produced a better definition of cytochrome absorption bands because of increased light scattering. Furthermore, Keilin showed that the low dispersion microspectroscope gave better visualization of cell and tissue absorption bands than did a higher resolution spectrograph.

Almost "across the street" at Cambridge University (i.e. in the Physiology Department diagonally from the Molteno Institute), a completely different approach following the classical ideas of Tyndall was being tried. He used a differential spectrophotometer in the infrared region (Figure 1)[3]. Tyndall employed a reference radiation source, a sample radiation source and a differential galvanometer. He balanced one against the other and achieved highly sensitive absorbance detection. Millikan modified the Tyndall spectrophotometer by using a photovoltaic cell divided into two parts along the mid-line and appropriate red and green filters* over each part for differential detection of the absorbance changes of hemoglobin and myoglobin in the cat soleus muscle[4].

---

* Thus Millikan may well have been using near infrared wavelengths "NIR" of Butler and Norris[5], and Jobsis[6] even in 1935.

Photon Migration in Tissues
Edited by B. Chance
Plenum Press, New York

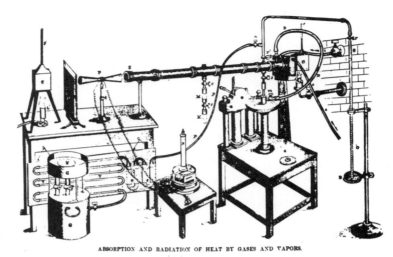

ABSORPTION AND RADIATION OF HEAT BY GASES AND VAPORS.

Fig.1.   Tyndall's dual wavelength   spectrophotometer[3]: Two infrared radiation sources are raised
to appropriate temperatures by Bunsen Burners.   The reference source radiation (e)
strikes the screen and the measuring source radiation (a) passes through tube S which
contains an absorbing liquid.   A differential thermocouple, P, measures the transmitted
infrared light and a differential galvanometer, N, measures this difference current.   The
system can initially be balanced to give a null rating on the galvanometer by adjusting the
intensity of the sample or the reference source.   Admission of a gas or liquid into the
sample tube will then cause a displacement of the galvanometer that is proportional to its
absorption of infrared radiation.

Curiously enough, at the same time, sophisticated theories concerning the diffusion of light in
highly scattering objects were developing, for example those of Kubelka and Munk[7], without
apparent impact upon each other.

While other work on tissue spectroscopy lay fallow in England and the U.S.A. during
World War II, Keilin and Hartree[8] developed low temperature microspectroscopy of cells and
showed advantages due to the sharpening and intensification of cytochrome absorption because
of the presence of scattering material and freezing to 77°K.   Also, Millikan and
Pappenheimer applied the split detector and red-green filter system to the ear lobe to make
the Millikan oximeter, which still serves as a reliable indicator of peripheral hemoglobin
oxygenation[9,10].

However, during World War II, research progressed rapidly in Sweden, and H.
Lundegårdh began his study of single beam scanning of root bundles[11]. Lou Duysens published
his dissertation in 1951 on light scattering and absorption in plant leaves and cells[12]. At
about the same time, Slater and Holton (1953) employed the single beam method for studies

of cytochromes in suspensions of mitochondria[13]. Chance initiated a concerted effort to study cytochromes and to improve upon Keilin's and Millikan's method[14]. Chance observed significant artifacts with the single beam due to swelling of the organelles or settling of the cells. In order to test the dual wavelength principle, Chance placed the sample at the entrance slit of two monochromators (each observation was then made at slightly displaced wavelengths, one set at an isosbestic point and the other set at a nearby absorption maximum, for example at 540/550 nm for studies of the oxido-reduction state of cytochrome c[15,16]). This method, using f=9 quartz monochromators (Beckman DU) was as optically inefficient as it might have been expected to be; the small numerical aperture of the monochromators failed to gather adequate light from the highly scattering biological materials. To improve light gathering, the system was redesigned so that the sample was placed next to the detector. With this arrangement a large angle of light was collected. The idea of synchronized chopping of the output from the two monochromators allowed the rapid switching of the incident light from one monochromator to the other so that synchronous detection by a lock-in type of detector was possible[17].

The need for the proximity of sample and detector was verified in studies by W. Slater who studied scattering and diffraction of light by muscle[13]. Since a large cathode cadmium-antimony photodetector was available, a relatively large (4 sq. cm.) cross-section sampling area was possible with near optimum optical geometry. The weak point of the system was the Beckman amplifier, which was similar to the Dubridge electrometer (but was dynamically unstable). Despite this, dual-wavelength recordings obtained from yeast cells[15,16] and suspensions of mitochondria were well delineated and scattering effects largely removed. The performance was subsequently improved upon with the Bausch & Lomb half-meter and quarter-meter grating monochromators, large diameter (2 inch) end-on photomultipliers and wide band operation[18].

In 1957, the spectroscopic studies of cytochromes in tissues, begun by Chance and Connelly, were confirmed and extended by Jobsis and Weber using the dual wavelength technique. Absorption changes related to mitochondrial function during muscle contraction were observed at wavelengths appropriate to cytochrome b and pyridine nucleotide[18-20]. All studies were carried out in hemoglobin-free amphibian muscle, usually the sartorious muscle.

In 1958, Warburg's early observation of the fluorescence of NADH (DPNH) *in vitro* was confirmed in *in vivo* studies of suspensions of yeast cells by Duysens[21] and in suspensions of mitochondria by Chance, who showed that the fluorescence signal originated largely from bound NADH of the mitochondrial matrix, a conclusion confirmed by microfluorometry of a single mitochondrial aggregate (nebenkern) in a single spermatid[22]. These observations diverted attention from cytochrome spectroscopy since the NADH signal showed a larger response, i.e. over 50% variation in intensity with altered metabolic function. Furthermore, with NADH, microspectroscopic observation of the brain lying between major vessels containing hemoglobin was possible. A combined hemoglobin spectrophotometer and NADH fluorometer was used to study rat brain hypoxia[23]. These studies clearly showed distinct responses of hemoglobin and NADH during hypoxia that were consistent with that expected *in vitro*: a low and a high affinity of hemoglobin and NADH for oxygen, respectively. Unfortunately, NADH fluorometry was limited to exposed tissue surfaces or internal surfaces that could be reached by catheterization.

Butler and Norris pioneered the near red region and applied the technique to plant tissues and to the human hand (5). Jobsis moved attention away from NADH tissue spectroscopy with his studies delineating "transcranial spectroscopy" of cytochrome oxidase[6,24]. He used the 830 nm absorption band of the copper component of cytochrome oxidase and sought, by use of various algorithms, to compensate for the absorbance changes of hemoglobin in the red region of the spectrum. In order to quantify hemoglobin and cytochrome concentrations four wavelengths were required. These were strategically located in the 700-900 nm region. However, the method was not capable of or indeed designed to cope with changes in tissue scattering that might occur in response to metabolic stress. Furthermore, the algorithms for evaluating cytochrome changes in human neonate brain were developed using perfused cat brain. The assumption of transferability, seemingly valid at the time, has been questioned by time-resolved spectroscopy of photon migration through cat and human brain studies, as shown below[25,26]. Thus the deconvolution of hemoglobin from cytochrome changes has undergone a difficult evolution. For this reason, the literature contains anomalous conclusions such as the concept that cytochrome oxidase is largely reduced (~50%) despite aerobiosis (FiO_2 = 20%). This has been attributed to a special form of cytochrome oxidase existing exclusively in brain tissue which does not saturate even at the highest tissue oxygen concentrations obtainable (for example, with inspiration of 100% oxygen and 10% CO_2). However, observations made on gerbil brain at low temperatures[27], where the sharpening and enhancement of cytochromes *c* and *a* absorption afford a better delineation of cytochrome and hemoglobin, showed no detectable cytochrome reduction during aerobiosis. In fact, no cyto-chrome reduction was detected until very low FiO_2's were reached and hemoglobin deoxy-genation was nearly complete. The conclusion was verified by Lubbers (pers. commun.) who found that no more than a 16% reduction of cytochrome *a* in the normoxic rat brain. Thus improved methods of dealing with living tissue light scattering seemed desirable.

TIME RESOLVED SPECTROSCOPY

The use of model systems for the study of photon migration has been exploited in a number of laboratories with great effectiveness (for example see Delpy, Bonner, Wilson, Blumberg[28-31] among others[6,32]). Our particular contribution has been the use of a larger scale model which better simulates the long path of photon migration in the human head. Furthermore, the specific goal of the detection and quantitation of hemoglobin deoxygenation has been emphasized. The model shown in Figure 2 provides a scattering layer (artificial milk was initially used, now yeast cells are preferred) of variable thickness which simulates the relatively low hemoglobin content of the skull and allows for "metabolically active" deoxygenation by metabolizing yeast and blood[25]. Thus the signal generated is not static but instead provides a linear decrease of oxyhemoglobin concentration with time. We are able therefore to simulate precisely the changes that would be expected to occur *in vivo* during hypoxic stress.

Results summarized elsewhere[33] give profiles developed by separating light input and output photons and varying the thickness of the layer between the photons and the milk itself. Subsequent to our initial publication[25], this model and another similar one have been employed together with the newly developed technique of time-resolved spectroscopy to quantify further the effects of increasing hemoglobin concentration . As shown, for example, in Figure 3, the presence of hemoglobin markedly changes the kinetics of photon exit.

Test System for TCSPC

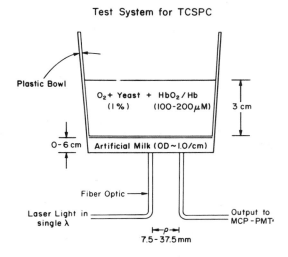

Fig. 2. Simulation of skull brain relationship.

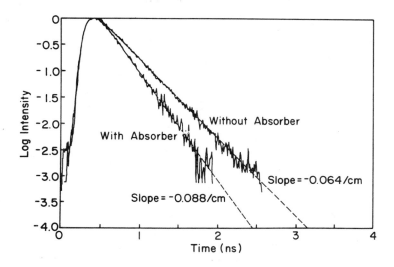

Fig. 3. Photon migration in milk model with and without Hb.

A Cat Head Model of Photon Migration

Time resolved spectroscopy was also applied to a cat brain and skull from which the hemoglobin had previously been removed by saline perfusion. The photon propagation characteristics of this cat head are shown in the left hand portion of Figure 4 as linear intensity and in the right hand portion as a logarithmic scale[25]. A strictly logarithmic function is shown rather than a power law function, due to the absorption of residual hemoglobin, cytochrome, etc. We have not yet observed the theoretical power law decay in tissues that are virtually blood free[29], as is observed in models.

Hemoglobin injection into one hemisphere of the blood free cat brain via an internal carotid artery is an imperfect model because of postmortem changes in the blood brain barrier. However, the injected hemoglobin produced an increase of 0.05 cm^{-1} in the log slope (Fig.4, panel B). The increment of hemoglobin concentration increase was calculated using $\Delta C = (-1/\Delta\varepsilon L) \log I$ to be $\Delta\varepsilon = 0.25$ cm^{-1}mM^{-1}. This corresponds to a hemoglobin concentration of 52 and 48 μM for Figures 4B and 4D. This value for hemoglobin was reasonably close to a total tissue hemoglobin assay by spectroscopy of 50 μM. Thus, a key advantage of these pulsed wave techniques is the ability to detect the time of photon migration and thus path length, solving one of the major problems of spectroscopy in scattering media.

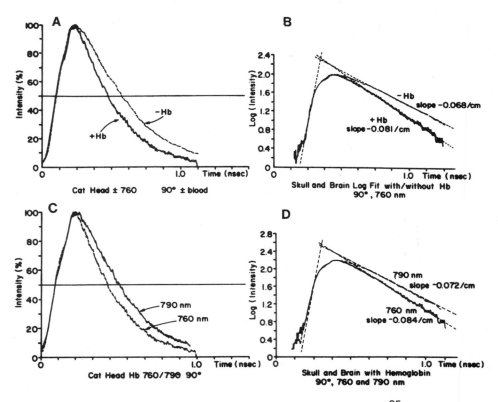

Fig. 4. Photon migration patterns in the perfused cat head[25].

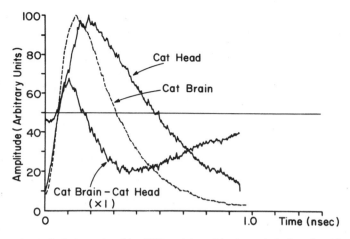

Fig. 5. Comparison of photon migration (790 nm) in cat brain and skull and cat brain alone.

One of the more important observations from these studies is shown in Figure 5. Here the propagation of light is shown to be much poorer in cat brain without skull than in the skull encased brain. For example at 790 nm, the hemoglobin free cat brain gave a value for photon decay of 0.117 cm^{-1} instead of 0.068 cm^{-1} for skull and brain or two times better propagation for the skull and brain. Since the exposed postmortem brain would be expected to expand due to osmotic effects and thus should be at least as large as the skull encased brain, the less effective light migration is attributable to the effect of boundary conditions. Any boundary that produces backscatter rather than an "escape" of photons appears to improve the propagation. Furthermore, the two studies seem to support the idea that absorption or backscatter from the skull prior to entry of light into the brain is not observed. In fact, a faster rise-time is observed from the exposed brain than from the intact brain, ensuring that the photon migration observed is from within the skull encased brain, as further supported by the hemoglobin injection study described earlier. Thus, short path migration of photons "trapped" within the skull and failing to enter the brain seems improbable.

Correlation of Time Resolved (TRS) and Continuous Wave (CWS) Spectroscopy

We have compared measurements of hemoglobin absorbance using our TRS with CWS. The weak point of CWS observations is the inability to determine optical path length, while the major asset of the TRS is that the path length is continuously measured. However, the TRS is considerably larger than a corresponding CWS unit thus its application, both

127

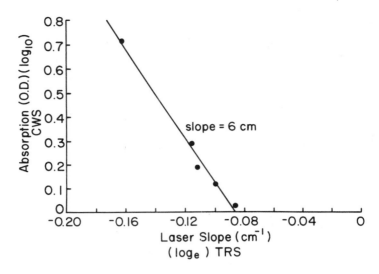

Fig. 6 Comparison of TRS and CWS in normally perfused dog head.

clinically and experimentally, is, at present, limited. However, the use of identical input-output geometries for CWS and TRS combined with a significant perturbation of hemoglobin concentration in an organ under study affords quantitation of the CWS output since the mean path length obtained from correlation plots of CWS and TRS allows a calculation of concentration change not only for the perturbations used for calibration but also during perturbations that might be expected under other conditions where TRS is not available.

To produce measurements of path length and absorbance in an organ as hemoglobin concentration is decreased, optical fibers for both TRS and CWS were fixed to the skull of an anesthetized dog and exchange transfusion with a fluorocarbon solution was carried out so that hemoglobin fell from its normal concentration to a low value of 1%. For these calibration studies a multi-diode array spectrometer capable of near simultaneous measurement from 600 to 1000 nm in 1 nm bands was used for CWS. At 760 nm (Figure 6) and normal hemoglobin concentration, both TRS and CWS showed a peak due to the normal amount of hemoglobin desaturation. After exchange transfusion, very little hemoglobin signal could be detected by CWS (Figure 7). However, the CWS baseline is very wavelength dependent in this nearly hemoglobin free brain, in contrast to the TRS signal which shows little change with wavelength. This suggests the feasibility of absolute determination of hemoglobin content with TRS. Note also that the cytochrome signal is too small to be observed at this sensitivity, thought it should have been present because the brain was well oxygenated by the fluorocarbon.

Figure 8 shows the relationship between TRS and CWS measurements of hemoglobin saturation during exchange transfusion with a progressive decrease in hemoglobin concentration. Note that the relationship is linear with a slope of about 6 cm which is the mean migration path for CWS through the dog brain for the particular fiber placement geometry used.

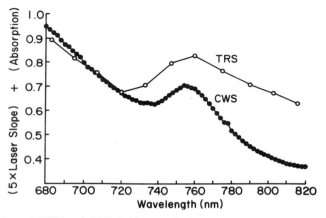

Fig. 7. Comparison of TRW and CWS for fluorocarbon perfused dog head residual Hematocrit of 1%

Use of CWS and TRS on the Adult Brain

The obvious application of this technology is to detect oxygen deficits in the human brain related to stroke, transient ischemic attacks, intracranial bleeding, cerebral aneurysms etc. CWS can be used conveniently in the operating room where crowding of instrumentation, surgeons and nurses about the patient allows no space for the application of more than a small transducer to the head or to a skeletal muscle, with necessarily long leads attached to remote instruments. Figure 9 illustrates the application of CWS, calibrated as above, to the forehead and the quadriceps muscle of a patient undergoing periodic cardiac arrest during implantation of an automatic defibrilator. The instrumental sensitivity for brain and muscle are identical. As seen during the approximately 30 sec interval of complete ischemia, the absorbance change due to hemoglobin deoxygenation in the brain is several times that of skeletal muscle.

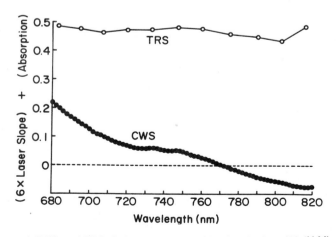

Fig. 8. Comparison of CWS and TRS during hematocrit reduction 40% to 1% (H.Miyake, unpub.obs).

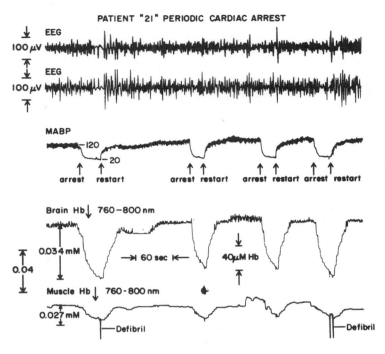

Fig. 9. Repetitive cardiac arrest and defibrilation in adult human head and quadriceps muscle.

TRS of the Brain

In order to quantify the path length of light in the adult human brain, the same TRS technique as employed in the dog model was used (Fig. 10). In this case, 630 nm was used. The exponential decay can be observed over four decades of intensity. When 760 nm is used, the slope (μ) is significantly smaller than that of the cat and dog brain (0.054 cm)[25]. Thus, the propagation of light in the larger adult human brain is more effective than in the smaller cat brain.

Figure 11 shows that the distance between input and output fibers has little effect upon the characteristics of the migrating photons; slopes (μ) are increased only slightly with 2, 4 and 6 cm fiber separations.

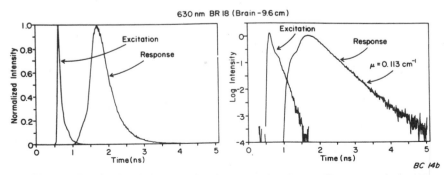

Fig. 10. Photon migration in adult human head measured at 9.6 cm fiber separation at 630 nm.

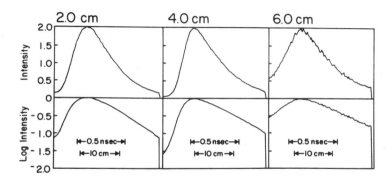

Fig. 11. Effect of fiber separation upon photon migration in the adult human head at 760 nm.

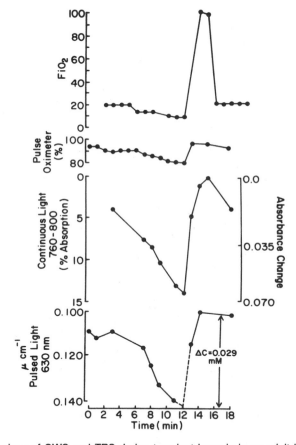

Fig. 12. Comparison of CWS and TRS during transient hypoxia in an adult human volunteer.

Comparison of TRS and CWS in the Adult Human Brain

The allowable perturbations of hemoglobin oxygenation are limited with human subjects. However, volunteers breathing hypoxic gas mixtures have provided correlations of TRS with CWS. Figure 12 shows a typical recording of pulsed light and continuous light data indicate during hypoxia and restoration of oxygen. A number of features are apparent

131

from this recording. The first and most important is the difference between finger pulse oximetry and brain hemoglobin absorbance as measured with both CWS and TRS The pulse oximeter indicates only a slight decrease in peripheral oxygenation, whilst the CWS and TRS brain monitoring indicate rapid depletion of oxygen.

The correlation plot of CWS and TRS for conditions where both measurements were varying rapidly in the progression towards anoxia is plotted in Figure 13 The slope allows calibration of CWS as follows:

$$\Delta C_{TRS} = \frac{\Delta \mu_{630}}{\Delta \varepsilon_{630}} = \frac{0.030}{1.37} = 0.022 \text{ mM}$$

Since ΔC's are the same

$$L_{CWS} = \frac{\Delta OD_{760-800}}{\Delta C_{TRS} \, \Delta \varepsilon_{760-800}} = \frac{0.048}{0.022 \times 0.25}$$

$$L_{CWS} = 8.7 \text{ cm}$$

It is in this way that the CWS method may be converted from a trend indicator to a quantitative spectroscopic technique for a particular set of conditions.

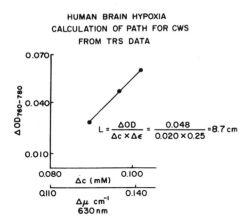

HUMAN BRAIN HYPOXIA
CALCULATION OF PATH FOR CWS
FROM TRS DATA

$$L = \frac{\Delta OD}{\Delta c \times \Delta \varepsilon} = \frac{0.048}{0.020 \times 0.25} = 8.7 \text{ cm}$$

Fig. 13. Correlation of CWS and TRS for frontal region of adult human head.

PROSPECT

While these studies have focused upon the responses of brain, it is obvious that many other oxygenation problems, can be studied by TRS, such as muscle oxygenation[26]. Last but not least, the development of appropriate algorithms for imaging ischemic/hypoxic tissue volumes are possible but depend upon the fundamentals of light propagation as described here. It seems clear that migration efficiency is critically dependent upon the location of the light absorber in a tissue. Thus reconstruction of the location of such absorbers is determined by alterations of photon migration in and out of a variety of locations near the absorber.

SUMMARY

The use of pulsed light for determining the length of the optical path and studies from animal models and humans indicates that:

1) propagation of light in the human brain is highly efficient as indicated by the value 0.05/cm for hemoglobin containing brain, and this value is a consequence of a wide ranging migration of photons in the brain;

2) the observation of photon migration pathways of nearly a meter confirms the efficiency of the migration process;

3) of most practical value is the fact that the path length of migration for continuous light, which is the mean of many optical paths, can be determined by comparing absorbance changes using both techniques in similar geometries and comparable positions on an animal model stressed to cause changes of the oxygenation state of hemoglobin.

ACKNOWLEDGEMENTS

This work was supported in part by National Institutes of Health NS 27346, HL-07286 (DSS), RR 01348 (GH), NS 23859 and NIM, Inc. and Otsuka Electronics, USA.

REFERENCES

1. D. Keilin and E. F. Hartree, Cytochrome and cytochrome oxidase. *Proc. Roy. Soc. London B.* 127:167-191 (1939).

2. D. Keilin, "The History of Cell Respiration and Cytochrome." Cambridge University Press, Cambridge, England, 1966.

3. J. Tyndall, "Contributions to Molecular Physics in the Domain of Radiant Heat." Appleton & Co., NY. Appleton & Co., New York. (1873).

4. G. A. Millikan, Experiments on muscle haemoglobin, *Proc. Roy. Soc. London B.* 123:218- (1937).

5. W.L. Butler and K.H. Norris. The spectrophotometry of dense light-scattering material. *Arch. Biochem. Biophys.* 87:31-40 (1960). See also Smith, K.C. ed. "The Science of Photobiology" Plenum Press, New York, p.400, 1977.

6. Jobsis, F.F. Non-invasive, infra-red monitoring of cerebral and myocardial oxygen sufficiency and circulatory parameters. *Science* 198:1264-1267 (1977).

7. P. Kubelka and F. Munk, Ein beitrang zur optik der farbanstriche. Zeits. f. tech. Physik, 12:593 (1931); see also Kubelka, P. New contributions to the optics of light scattering materials. Part I. J. Opt. Soc. Am 38:448 (1948); and Part II. J. Opt. Soc. Am 44:330 (1954)

8. D. Keilin and E.F. Hartree, Effect of low temperature on the absorption spectra of haemoproteins; with observations on the absorption spectrum of oxygen. *Nature London* 164:254 (1949).

9. G. A. Millikan, The oximeter, an instrument for measuring continuously the oxygen saturation of arterial blood in man. *Rev. Sci. Instrum.* 13:434 (1941).

10. J. R. Pappenheimer. Vasoconstrictor nerves and oxygen consumption in the isolated perfused hindlimb muscles of the dog. *J. Physiol.* 99:184 (1941).

11. H. Lundegårdh, Action spectra of the reducing and oxidizing systems in spinach chloroplasts. *Biochim. Biophys. Acta* 88:37 (1964).

12. L.M.N. Duysens Thesis, Utrect 1951; see also L.M.N. Duysens, Photosynthesis. In "Progress in Biophysics in Molecular Biology", Pergamon Press, 1964..

13. E. C. Slater and F.A. Holton, Oxidative phosphorylation coupled with the oxidation of a-ketoglutarate by heart-muscle sarcosomes. I. Kinetics of the oxidative phosphorylation reaction and adenine nucleotide specificity. *Biochem. J.* 55:530 (1953).

14. B. Chance, Rapid and sensitive spectrophotometry III. A double beam apparatus. *Rev. Sci. Instrum.* 22:634 (1951).

15. B. Chance, Enzymes in action in living cells: The steady state of reduced pyridine nucleotide. *The Harvey Lecture Series* 49:145 (1955).

16. B. Chance, Enzyme mechanisms in living cells. in: "The Mechanisms of Enzyme Reaction" W. D. McElroy and B. Glass, eds., The Johns Hopkins Press, Baltimore, pp. 399- 453 (1954).

17. B. Chance, Rapid readout from dual wavelength spectrophotometer. *Rev. Sci. Instrum.* 43:62 (1972).

18. C. M. Connelly and B. Chance, A sensitive spectrophotometric method for reading enzyme reactions in cell suspensions, muscle and nerve. *Am. Philos. Soc.* 227:710 (1954).

19. B. Chance and F. Jobsis, Changes in fluorescence in a frog sartorius muscle following a twitch. *Nature* 184:195 (1959).

20. B. Chance and A. Weber, The steady state of cytochrome b during rest and after contraction in frog sartorius. *J. Physiol.* 169:263 (1963).

21. L. M. N. Duysens, The flattening of the absorption spectrum of suspensions, as compared to that of solutions. *Biochim. Biophys. Acta* 19:1 (1956)

22. B. Chance, G. Mauriello, and X.M. Aubert. ADP arrival at muscle following a twitch. In "Muscle as a Tissue" (Rodahl, K. and Horvath, S.M., eds.) McGraw-Hill Publishers, New York, pp. 128-145 (1961)

23. B. Chance, N. Graham and D. Mayer. A time sharing fluorometry for the read out of intracellular oxidation-reduction states of NADH and flavoprotein. *Rev. Sci. Instru.* 42:951-957 (1971).

24. H.J. Proctor, A.L. Sylvia and F.F. Jobsis. Failure of brain cytochrome aa_3 redox recovery after hypoxic hypotension as determined by *in vivo* reflectance spectrophotometry. *Stroke* 13:89 (1982).

25. B. Chance, J.S. Leigh, H. Miyake, D.S Smith, S. Nioka, R. Greenfeld, M. Finander, K. Kaufmann, W.Levy, M. Young, P. Cohen, H. Yoshioka,and R. Boretsky, Comparison of Time Resolved and Unresolved Measurements of Deoxyhemoglobin in Brain. *Proc. Natl. Acad. Sci. USA* 85:4971-4975 (1988)

26. B. Chance, S. Nioka, J. Kent, K. McCully, M. Fountain, R. Greenfeld,and G. Holtom, Time Resolved Spectroscopy of Hemoglobin and Myoglobin Resting and Ischemic Muscle. *Anal. Biochem.* 174:698-707 (1988).

27. C.L. Bashford, C.H. Barlow, B. Chance and J. Haselgrove. The oxidation-reduction state of cytochrome oxidase in freeze trapped gerbil brains. *FEBS Lett.* 113:78-80 (1980).

28. P. vanderZee and D.T. Delpy. Computed-point spread functions for light in tissue using a measure volume-scattering function. Intl. Soc. Oxygen Transport to Tisssue (ISOTT) (July 22-25), Sapporo, Japan, p. 82 (1987); see also: P. vanderZee and D.T. Delpy. Simulation of the point spread function for light in tissue by a Monte Carlo technique. *Adv. Exp. Med. Biol.* 215:179-192.

29. R. F. Bonner, R. Nossal, S. Havlin and G.H. Weiss, Model for photon migration in turbid biological media. *J. Opt. Soc. Am. Sec.* A4:423 (1987)

30. B. C. Wilson, M. Patterson, S.T. Flock and J.D. Moulton. The optical absorption and scattering of tissues in the visible and near-infrared wavelength range. In "Light in Biology and Medicine" (Douglas, R.H., Moan, J. and dal'Acqua, eds.) Plenum Publ., New York.

31. W.E. Blumberg, Light propagation in human tissues: The physical origin of the inhomogeneous scattering mechanism. *Biophys. J.* 51:288 (1987).

32. O. Hazeki and M. Tamura. The quantitation analysis of hemoglobin oxygenation state of rat brain in situ as monitored by near infra-red spectroscopy. *J. Appl. Physiol.* 64: 796-802 (1988).

35. Greenfield, R., M.S. Thesis, University of Pennsylvania, Philadelphia, PA 19104

Part 3.   Instrumentation Aspects

EXPERIMENTAL TIME RESOLVED METHODS

Gary R. Holtom

Regional Laser and Biotechnology Laboratories
University of Pennsylvania

INTRODUCTION

Recent experiments involving light scattering and diffusion in tissues show a substantial signal emerging from large tissue masses on the order of 2-10 nanoseconds (ns) following light entry. A quantitative method for resolving the scattered light should have these attributes:

a) High sensitivity. This implies a quantum limited detector such as a photomultiplier tube (PMT).
b) Good dynamic range, ideally five decades if needed.
c) Stable, linear, and adjustable time sweep for convenient comparison of data to theoretical models.

Fortunately these requirements are the same as needed for fluorescence decay measurements, for which the technology is very highly developed. This article will describe one method, time-correlated single-photon counting (TCSPC), and indicate its applicability to tissue scattering experiments. The use of pulsed lasers and computer assisted data acquisition is a natural part of this method.

It is worth mentioning that several more obvious methods are not as appropriate. Direct measurement on an oscilloscope has dynamic range and sweep linearity problems. Obtaining a background-free signal with large dynamic range is difficult with single-pulse measurements and recording the detector signal directly. The streak camera has good picosecond response but is not well suited for long time sweeps. The ultrafast methods have the very low sensitivity characterized by optical gating which replaces the use of fast detectors.

Photon Migration in Tissues
Edited by B. Chance
Plenum Press, New York

In the last few years, phase modulation methods have been developed which are very competitive in time resolution with the TCSPC method described here. There are advantages and drawbacks. A CW source can be modulated inexpensively, but this becomes increasingly difficult at very high frequencies, and the best phase systems use a cavity dumped dye laser like that described here. The detector becomes a problem in the GH_2 domain since the anode pulse width is more difficult to control than its transit time jitter. Finally, data analysis is less intuitive since data is collected in the frequency domain rather than the time domain.

Nonetheless, phase methods may well find a place for routine experiments where the very best resolution can be sacrificed for simplicity.

THE TCSPC METHOD

Flashlamps were used as light sources for TCSPC before short pulse lasers became commonly available[1]. The cavity dumped dye laser has been used as an improved light source for more than a decade[2], and there have been continuous improvements in the technique since then[3]. An experimental outline is presented in Figure 1.

It is useful to think of the experiment as a series of very simple measurements. A pulse of light is injected into the sample and simultaneously triggers a time zero marker diode

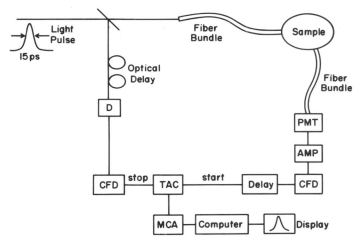

Fig. 1. Detection and timing electronics for TCSPC. Part of the light pulse reaches the trigger diode D after delay through the optical fiber coil. The remainder illuminates the sample, and migrating photons are received by the photomultiplier tube PMT. Analog signals from both D and PMT amplifier AMP are converted into fast logic pulses by the constant fraction discriminator CFD, and the time difference between them is recorded by the time-to-amplitude converter TAC. The multichannel analyzer MCA digitizes and stores the signals for viewing and processing by a small computer.

D. At some point one of the emitted photons makes its way to the PMT and the discriminator CFD produces a timing pulse. The object is to measure the time delay between the timing pulses from the diode and PMT and this is done by the time-to-amplitude converter (TAC). This process is repeated until a total of 10^5 to 10^8 pulses are accumulated, depending on the noise level required to achieve the desired accuracy. Data collection times are typically 20 to 200 s.

Fortunately very convenient nuclear instrumentation modules exist for processing the signals. The TAC starts an analog sweep following a start pulse which is held at constant voltage upon receipt of a stop pulse. This voltage is equivalent to a time axis, and by means of analog amplification the time sweep can be varied over a wide range. For a few microseconds the voltage is held constant, then reset. This pulse is digitized and increments one of several hundred memory channels in the multichannel analyzer MCA, which has memory shared with the data processing computer. While data is being collected, the various channels corresponding to different times contain increasing counts, which are visible on the computer screen.

Since it is not convenient to change the optical delay time (fiber length), it is made longer than necessary and a switchable electronic delay is provided for the PMT timing pulse to accommodate changes in sample position. After a number of pulses have accumulated the MCA contents are a record of the scatter times for the photons. It is important to set the detected count rate to a small fraction of the laser pulse rate (typically 1%) to ensure that late photons are detected with a probability equal to early ones. Most MCAs are limited to a conversion rate of 40 KHz which just happens to be two orders of magnitude slower than the usual laser rate of 3.8 MHz.

An explanation of the TAC operation is necessary. Laser pulses are typically provided at intervals of 250 ns, which is much shorter than the reset time of the TAC. For this reason, the TAC is not started when the light pulse enters the sample, but rather when a photon is detected. In effect, this avoids dead time and reverses the time axis. It is necessary to delay the diode pulse, now the stop pulse, by the sweep time of the TAC. This is conveniently done by a fiber optic coil rather than a long electrical cable which better preserves the short rise time of the timing signal.

THE LASER

The cavity-dumped dye laser is a very flexible light source (Figure 2). The pulses are short, on the order of 10 ps for routine operation. A typical repetition rate is 3.8 MHz but this can be varied as needed. While a high repetition rate is essential for prompt data collection, mode locked lasers with pulse separations of 10 to 13 ns (76 to 100 MHz) are not generally used since the timing electronics require a longer reset time, and confusion can

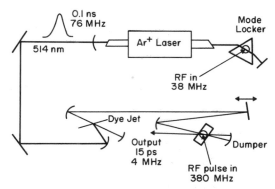

Fig. 2. Pulsed laser design. Green light is produced with an argon ion laser containing an acousto optic modulator, making a continuous train of mode-locked pulses. These are converted to tunable red light by the dye laser, whose cavity length is matched to the argon laser. All of the mirrors are high reflectors, which trap and store the light in the cavity. On demand a short radio frequency burst is sent to the cavity dumper to eject one light pulse and direct it to the experiment.

arise during data analysis if light is still being collected from one laser pulse when the next one arrives.

Dye lasers are the most common mode locked and cavity dumped laser, and are most useful as research tools due to their tunability. They are most often excited with green light from a pump laser, allowing light from 550 to 900 nm to be generated routinely. While several dyes are required to cover this range, each dye can cover as much as 100 nm and this is very convenient for performing spectroscopic measurements.

A pump laser which can be mode-locked is necessary for the cavity-dumped dye laser. The laser is in effect modulated at high frequency to generate a continuous train of pulses. Common sources are the argon ion laser at 514 nm and the second harmonic of the Nd:YAG laser at 532 nm. These provide about 1 W of green light with a pulse width of about 0.1 ns and are very satisfactory for TCSPC.

If shorter wavelengths are desired, the dye laser can be frequency doubled with an efficiency of several percent since the peak power is in the kilowatt range. This near ultraviolet and blue light is more important for fluorescence lifetime measurements than for tissue scattering.

COLLECTING THE SIGNAL

The tissue under study is irradiated with the laser beam at a suitable spot. Since the subject may be awkward to move and to maintain at an absolutely fixed location it is convenient to use a fiber optic bundle for beam delivery. It is important to remember that light requires three picoseconds (ps) to move one mm, and if the optical path length to the sample changes by even a few millimeters the absolute time zero is lost. It is equally appropriate to use a fiber optic bundle to collect the scattered light and transmit it to the PMT. This permits considerable flexibility in working with a subject.

Some control over signal level is possible by selecting the size of the fiber optic bundles. The 3 mm size is most frequently used but could be increased for harvesting more light from long scattering paths. It may also be useful to use much smaller fibers instead of attenuators for strong signals.

THE DETECTOR AND INSTRUMENTAL RESPONSE FUNCTION

Conventional fast PMTs suitable for photon counting are end-window (such as the XP-2020) or side window (such as the 1P-28). The best of these produce an instrumental response function of 0.2-0.5 ns FWHM, which is clearly not laser limited in width. The fastest PMTs are the microchannel plate type, with instrument response functions as short as 20 ps FWHM[3,4].

The instrumental response function (or instrument function) is obtained by removing the sample and attenuating the laser by 3-7 orders of magnitude, then allowing the measurement of a zero time delay signal (Figure 3). This includes all of the fixed delays and contributions toward the shape introduced by the laser, PMT, coupling fibers, and timing electronics.

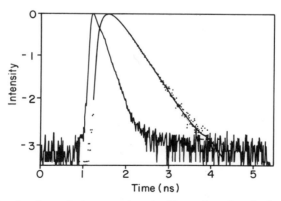

Fig. 3. Data from the thumb scattering experiment. The pulse at early time is produced by butting the two fiber optic bundles and is the instrumental response function and zero time marker. Scattered light shows up as dots at later time and the smooth curve is a convoluted response using the function given in the text. A logarithmic scale is used with about 10^4 counts peak.

In the most usual form of data analysis, a model for intensity versus time is assumed and is convoluted with the instrumental response function[1]. The result should agree to within a scale factor with the observed scatter signal, and statistical methods are useful to determine an objective quality-of-fit criterion. Systematic deviations between the data and calculated curves are used to predict changes in the model parameters which improve the fit, and a new curve is drawn with the "adjusted" parameters. This process, often called deconvolution, is more accurately known as iterative reconvolution[5].

With modern electronics the dominant contribution to the instrument function width is the spread in transit time experienced by electrons as they propagate from the photocathode to the anode. The microchannel plate tubes have planar geometry and very short internal distances, which make the response insensitive to photocathode illumination position and the wavelength of light, which are substantial problems with conventional PMTs.

AN EXAMPLE OF DATA AND CURVE FITTING

For demonstration purposes the scattering of light through a thumb were recorded Figure 3 illustrates the most important features of collection and modeling of data. The dy laser wavelength was 790 nm. At this wavelength the bone is weakly scattering and there ar no strongly absorbing pigments, so that the dominant factor in producing a late photon (on which was trapped in the tissue for a long time) is the multiple scattering process. Both th instrument function (left curve) and the scattered signal (dots to the right) contain about 10 counts full scale and are presented on a logarithmic scale to show the dynamic range that i possible. Since the data is collected with a spacing of about 18 mm between the fiber bundle there is a time delay before the arrival of scattered signal.

The calculated scattered curve (smooth curve to the right) is reasonably well fit with the function

$$S_{(t)} = -e^{(-t/87 \ ps)} + e^{(-t/251 \ ps)},$$

where the first term is an exponential rise in the signal and the second is a slower decay. Included in this fit is a delay term of 90 ps which corresponds to the time required for the most direct photon to go from one fiber to the other.

MULTIPLE SITE COLLECTION

It is possible to segment the anode, each portion of which sees a small part of the photocathode. This makes it possible to use one detector with a large number of collection fiber bundles. It appears that simultaneous data collection from 16 or more sites is feasible using a single PMT and multiple channels of timing electronics (Figure 4). Presently analog

144

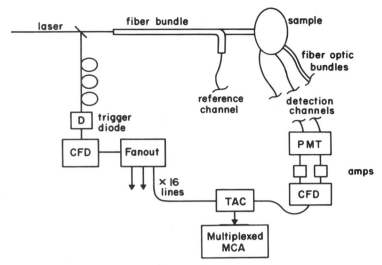

Fig. 4.   Strategy for simultaneous collection from multiple sites.   The image from several fiber
bundles is relayed to the PMT photocathode by a lens.   Each amplified current pulse is
collected on one anode segment and sent to one of the timing channels, and digitized with one
or more multiplexed MCAs.

timing channels consisting of amplifiers, CFD channels and TACs are much cheaper than the fast
microchannel plate PMTs and lasers.

Collection of multiple points simultaneously obviously reduces the time required to
perform multiple measurements.   Equally important is the possibility of getting relative
scattering amplitudes accurately at different spots even while the laser power has short or
long term fluctuations.

SPECIAL PROBLEMS

There are several problems associated with working in the near infrared (NIR) region
that do not appear in more conventional applications of TCSPC to the visible or near
ultraviolet region.   Alignment of dye lasers is not as convenient, and frequently it is necessary
to operate with a red dye and then change mirrors one by one while maintaining lasing after
each substitution.   This requires dyes and mirrors with overlapping spectral regions.

The NIR dyes are not soluble in pure ethylene glycol and require the addition of
propylene carbonate, a good solvent.   The recently introduced styril dyes are stable and
efficient, and can be pumped with green rather than with red light.   However, they tend to
produce much wider pulses in cavity dumped operation than the shorter wavelength rhodamine
dyes, unless the pulse width of the pumping laser can be kept very short.

The microchannel PMT is a variant of the end-window type, and requires a semitransparent photocathode. This rules out the very efficient GaAs material, which is at least an order of magnitude more efficient at 800 nm than the extended red multialkali surface. The very red S1 cathode is quite noisy and insensitive, and is completely unusable unless cooled to substantially below 0°C.

Changing the laser intensity over many orders of magnitude is necessary as tissue depth changes and while instrument functions are being recorded. Usually this is done by inserting filters, which is awkward and introduces time shifts due to path length changes. A very wide range neutral density wheel of constant thickness would be convenient.

Finally, the entire cavity dumped dye laser system must be regarded as a problem when moving from a research laboratory to a clinical environment. Alternate means for generating pulsed, or at least modulated, light at appropriate wavelengths must be considered seriously. An interesting source which may replace the dye laser is the krypton ion laser, which can be mode-locked to produce short pulses. Wavelengths selectable by means of an in-cavity prism are 568, 647, 676, 752, and 799 nm. Rapid development of semiconductor lasers may result in very convenient sources that can be pulsed at useful regions in the deep red. Other lasers which are considered high repetition-rate, such as Q-switched CW-pumped NdiYAG lasers, excimer lasers and metal vapor lasers, are some orders of magnitude too low in pulse rate to be useful.

ACKNOWLEDGEMENT

The apparatus used for this work was funded by the National Science Foundation grant CHE84-00198 and by the National Institutes of Health grant RR-01348. The NIH also supports the RLBL as a user facility.

REFERENCES

1. Lewis, C., W. R. Ware, L. J. Doemeny and T. L. Nemzek, The measurement of short-lived fluorescence decay using the single photon counting method, *Rev. Sci. Instr.*, 44:107-114, (1973).
2. Spears, K. G., L. E. Cramer and L. D. Hoffland, Subnanosecond time-correlated photon counting with tunable lasers, *Rev. Sci. Instr.*, 49:255-262, (1978).
3. Kume, H., K. Nakatusgawa, S. Suzuki and D. Fatlowitz, Ultrafast microchannel plate photomultipliers, *Appl. Opt.*, 27:1170-1178, (1988).
4. D. Fatlowitz, personal communication.
5. O'Connor, D. V., W. R. Ware and J. C. Andre, Deconvolution of fluorescence decay curves, a critical comparison of techniques, *J. Phys. Chem.*, 83:1333, (1979).

A TISSUE MODEL FOR INVESTIGATING PHOTON MIGRATION IN TRANS-CRANIAL INFRARED IMAGING*

Robert Lloyd Greenfeld

University of Pennsylvania
The Moore School of Electrical Engineering
Philadelphia, PA

INTRODUCTION TO TCSPC

Single photon counting time correlation[1-3] is a method to measure a sample's properties such that the output intensity is time-correlated with the input. Knowing this enables the user to deconvolve the response in order to determine the imulse response for a sample. It is very useful in studies which measure lifetimes of sample fluorescence.

In the system used here, a liquid dye laser produces the low power near infrared (NIR) light. Instead of the Nd:YAG laser, the TCSPC system uses an Argon ion (A+) laser at 514 nm. It is mode-locked with an acousto-optic modulator at 38 MHz so that it puts out a train of 100 ps pulses every 13 ns (76 MHz) to pump the LDL as shown in Figure 1. This train goes through the dye jet which shortens the pulses tremendously; 10 to 15 ps easily, less with some effort. The dye laser is cavity-dumped to the output every 20 pulses using an RF modulator (which is synchronized with the mode-locker) giving a total repetition rate of 4 MHz. On the other end of the sample is the receiver circuitry. A part of the input light is taken off to a trigger diode which synchronizes the output with the input. The output photons go to a multi-channel plate photomultiplier tube (MCP-PMT)[4,5]. The constant fraction discriminator (CFD) lets the timing be insensitive to the amplitudes of the trigger pulse and the output from the sample. The two trigger pulses give the start and stop signals to a time-to-amplitude converter (TAC) which sends the information to the multi-channel

---

* This contribution is an excerpt from the author's Master's thesis, University of Pennsylvania, 1988.

analyzer (MCA). This system contains 8164 channels and into each channel one or no photons enter for each pulse (i.e. no more than one photon will be incident on the PMT for each pulse). However, over the course of data collection, each channel is integrated, and the total number of photons collected is stored as an integer (which is in digital form already since the photons are discrete). The MCA is compressed by a factor of 16 to 512 channels. Thus, for a full scale window of 5 ns, each channel is 10 ps and the data can be plotted on a computer screen. The size of the window can be changed easily (say to 10 ns) simply by changing the gain on the TAC.

Since only one photon or less is incident on the PMT at any one time, the system can be considered to be working as a Poisson process. In each channel then, the output indicates the probability that a photon exited at that time. Figure 2 is the input laser function. As with the streak camera, the laser function width is not 10 ps, but rather a probability distribution of photons. However, with the TCSPC system, statistical data can be easily determined since the process is Poisson. A comparison of the streak camera and TCSPC (PMT) laser functions is shown in Figure 3. The top shows the functions normalized to their respective maxima and a similar FWHM of 100-120 ps. The difference can be seen in the logarithmic plot where the camera function drops off much more sharply. The prolongation of the TCSPC function is due mostly to the slower electronics, but since the sample response is slow and does not begin at the same time as the input (as it does in fluorescence), this extra fall time is not critical.

EXPERIMENTS WITH TCSPC

Human

To get more of an idea of the propagation of light in human tissue , we took spectra from the brain, calf, and arm. In the calf and arm, at fiber separation distances of 42 and 48 nm respectively (shown in Figure 4), the signal lasts approximately 1.8-2.2 ns after the peak and we find that the slope (i.e. the absorption) for calf and arm are 0.074 and 0.093 cm-1 and the FWHM are 675 and 534 ps, respectively. These values can be compared to those for the brain at 35 mm separation (see Figure 5) of 0.059 cm-1 and 860 ps. The differences are due to the effect of the skull, which extends propagation and differences between skeletal muscle and brain tissue. It seems that the muscle is a much higher absorber and has different boundary conditions[6]. In addition, we can plot the values of absorption in the brain vs. fiber separation and see that we still get signals for separations up to 10 cm (Figure 6).

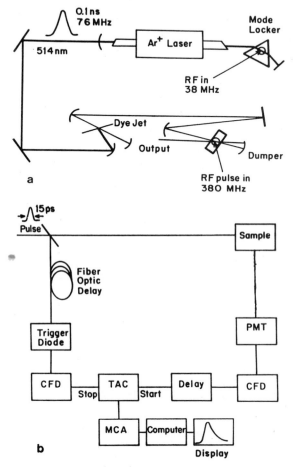

Fig. 1. Setup of time-correlated single photon counting system: (a) Laser diagram; (b) Electronics diagram.

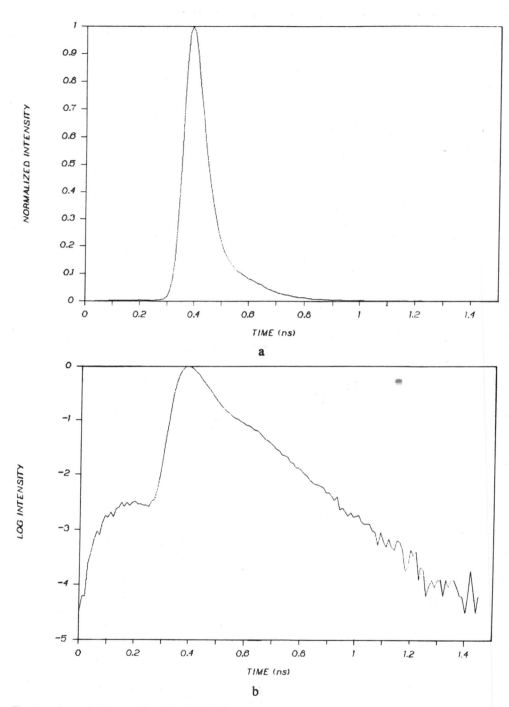

Fig. 2.   Laser instrument function for TCSPC system (on a scale of 0-2.5 ns): (a) normalized;(b) logarithmic.

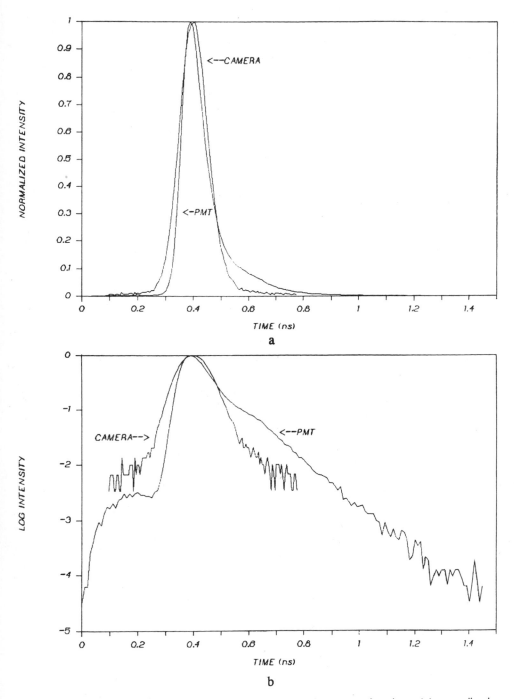

Fig. 3. Comparison of streak camera and TCSPC laster instrument functions: (a) normalized;
(b) logarithmic

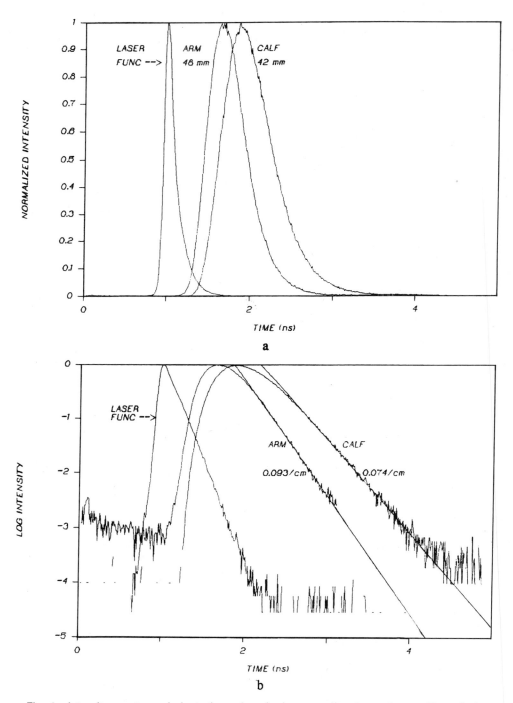

Fig. 4. Intensity spectra and absorption values for human calf and arm tissue with excitation
(instrument) function at 760 nm: (a) normalized; (b) logarithmic.

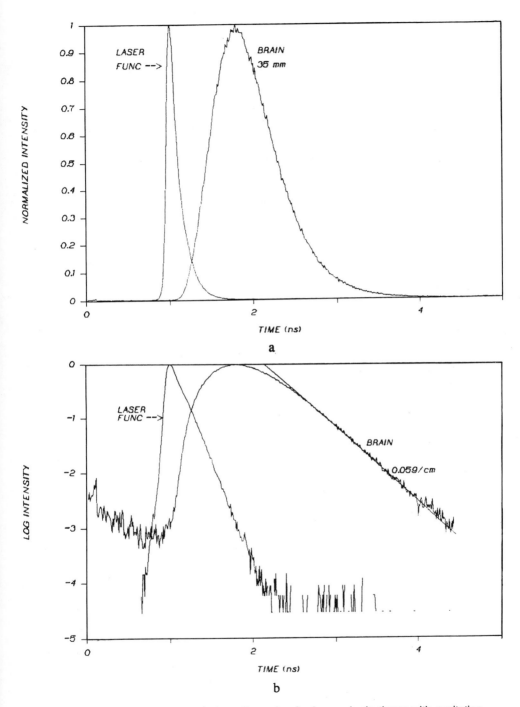

Fig. 5. Intensity spectra and absorption value for human brain tissue with excitation (instrument) function at 760: (a) normalized; (b) logarithmic.

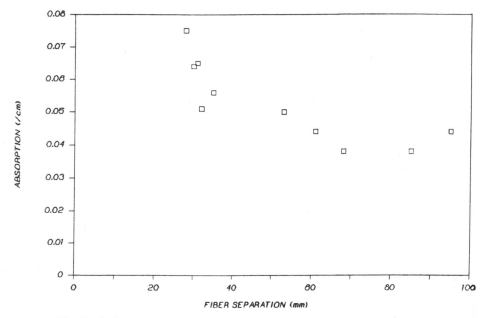

Fig. 6. Brain tissue absorption at 760 nm as a function of fiber separation.

Milk Model

Taking the brain tissue absorption as the one to mimic the milk model was used (7). The scatter material was an artificial milk solution with extinction ~ 1 OD/cm, and the absorber was HbO2. The setup was with one vessel which contained the scatterer. The source-detector setup changed because we only had one fiber for each. However (see Figure 7), the distance between source and detector could be varied from 7.5 mm to 37.5 mm (in increments of 7.5 mm), as compared to the fixed 20 mm separation with the rectangular light guide and the dual wavelength spectrophotometer.

With a fixed amount of scatterer in the bottom vessel (2.3 cm deep), we can see in Figure 8 that for smaller separations, the tail is not exponential, but rather biphasic or multiphasic. As the fiber separation approaches 30 nm, the response begins to straighten out and we start to see a single exponential response. The slopes (absorption values) of the "early" and "late" phases are given in Table I. Adding an absorber to the milk (same concentration as before) yields the spectra in Figure 9 at 30 mm. The absorber changes the slope by 30-40%.

Figure 8 introduces an interesting phenomenon. For a true scatterer, since it does not follow the Beer-Lambert Law, the tail should not be a single exponential. With the TCSPC system comes a program to deconvolve the output given the instrument function. The result is the impulse response of the system as a sum of weighted exponentials as follows:

$$h(t) = \sum_{i=1}^{4} A_i e^{-(t-to)/\tau_i}$$

(1)

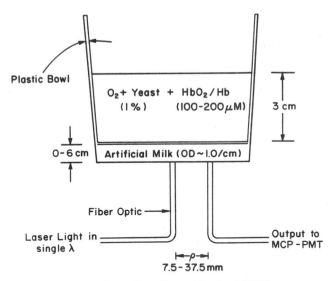

Fig. 7. Setup for milk model using TCSPC.

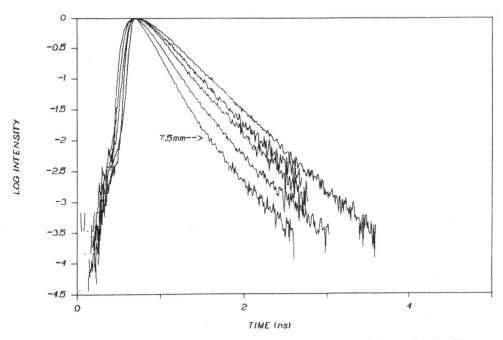

Fig. 8. Log intensity spectra of milk model showing non-exponential decays for five fiber
separations (7.5-37.5 mm). Peaks are normalized.

Table I. Milk model absorption values and measured characteristics lengths for difference fiber separations for spectra in Figure 8.

Separation (mm)	Absorption Value (cm⁻¹)			$L_{1/2}$ (cm)
	Early	Late	Single	meas.
7.5	0.1055	0.6092		3.57
15	0.0862	0.0644		4.95
22.5	0.0718	0.0583		6.56
30			0.0636	7.36
37.5			0.0574	8.51

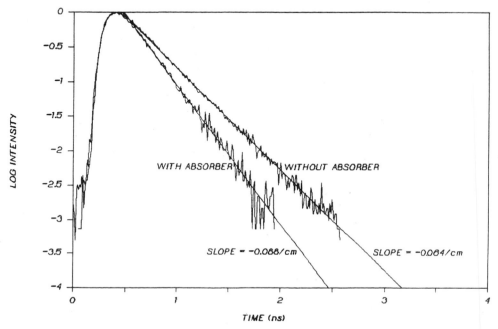

Fig. 9. Effect of absorber on the log intensity curve of the milk model at 30 mm.

For the milk model at a fiber separation of 15 mm, the impulse response is

$$h_{15}(t) = 0.053e^{-(t-0.072)/0.122} + 0.01e^{-(t-0.072)/0.303} \qquad (2)$$

The fitted curve is plotted versus the original curve in Figure 10; the χ squared loss is equal to 1.82. Bonner et al.[6] derive a theoretical intensity decay given an exponential absorber and isotropic scattering. As a function of time step n, fiber separation ρ, and absorption μ, the decay should be

$$\Gamma(n,\rho) = 0.055(n-2)^{-3/2}(1-e^{-6/(n-2)})(e^{\frac{-3\rho^2}{2(n-2)}-\mu n}) \qquad (3)$$

A typical plot of this equation for ρ = 15 mm and for values of L and μ (as given in ref. 6) of 0.23 cm and 0.048, respectively, is shown in Figure 11 along with the plot of $h_{15}(t)$. Even though Figure 11 looks like a response that obtained using the TCSPC system, we have to realize that it is an impulse response, not a convolution with a laser function. The TCSPC

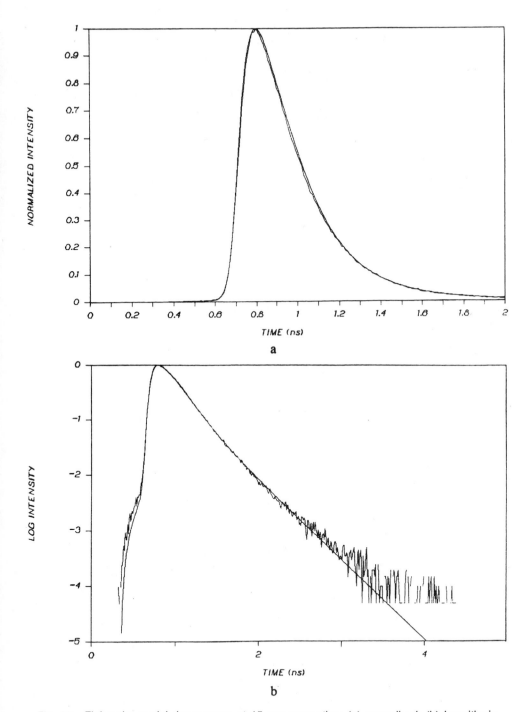

Fig. 10. Fitting the model decay curve at 15 mm separation: (a) normalized; (b) logarithmic.

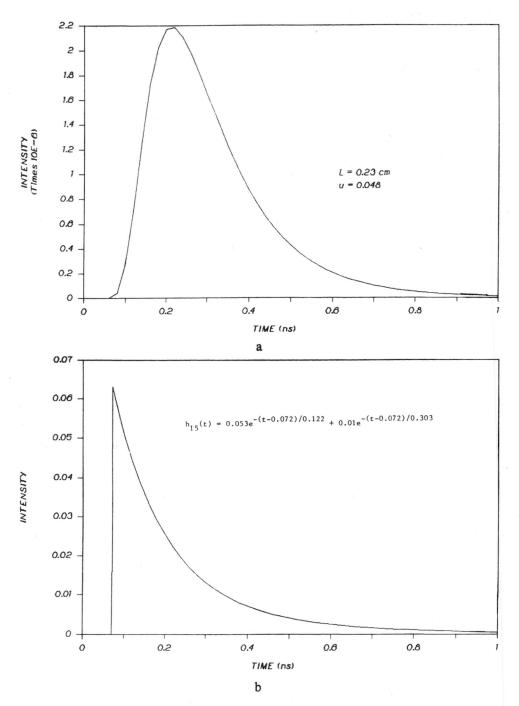

Fig. 11. Impulse response functions for 15 mm fiber separation: (a) Random-walk model: Eqn. 11; (b) Milk model, Eqn. 10.

impulse response does not begin at the origin, but is shifted 0.072 ns and from there it monotonically decreases over time.

ANIMAL MODELS

For a better understanding of the photon migration in vivo, to see how the absorption values change with wavelength, and to see how well the time-resolved method behaves regarding the Beer-Lambert Law, we set up a hypoxia study and a blood loss study.

Hypoxia

This experiment involved an anesthesized cat with the skin and fur on the head removed, but the skull intact. Figure 12 shows four light guides placed on the skull: an input and an output for the time-correlated system, and an input and an output for the scanning spectrophotometer. Each pair of fibers was separated by about 10 mm and were placed on opposite sides of the skull. The fraction of inspired oxygen (FiO_2) was varied using a mixture of oxygen and nitrogen gases. Besides taking data at each level of hypoxia, we also took data at a number of wavelengths in the near infrared regions.

The results are very impressive. Figure 13a shows the scanning spectrophotometer data of OD vs. time (FiO_2) for 760 and 802 nm during initial hypoxia and hyperoxia ($FiO_2 >$ normal). The 760 spectrum marks the increase in deoxyhemoglobin as the oxygen level decreases; the 802 spectrum shows the accompanying changes in blood volume and blood pressure (compensatory mechanisms to offset the oxygen changes). By subtracting the 802 data from those a 760, Figure 13b shows we can get a curve that depends only on the increase of Hb during hypoxia and the influx of HbO2 at 100%. Figure 14 shows the same data for the laser system - one can see the correlation with the FiO_2. Similarly, Figure 15 shows the wavelength dependence of the hypoxia for four values of FiO_2. If there were no blood volume changes or scattering, all the points at 802 nm would be about the same. Thus we see that the TCSPC system can give absorption data as well as a time-averaging spectrophotometer and can follow the Beer-Lambert Law. In fact, we can find the path length of the continuous wave (CW) system with a method introduced by Chance et al.[8,9]. Their equation relates the absorption (μ) of the time-resolved method (in units of cm^{-1}) to the OD of the CW method as follows:

$$\Delta OD_{CW:760-802} = \Delta \varepsilon c l_{CW} \qquad (4)$$

and

$$\frac{\Delta OD}{cm} = \Delta \mu = \Delta \varepsilon c \qquad (5)$$

therefore

$$\frac{\Delta OD_{CW:760\text{-}802}}{\Delta \mu} = I_{CW}(\text{pathlength}) \qquad (6)$$

For this hypoxia experiment, the data of ΔOD vs. $\Delta \mu$ using both FiO_2 and wavelength as parameters are shown in Figure 16. The slope of the line (~ 7 cm) is the average pathlength of the continuous system and is much more than the 1 cm fiber separation.

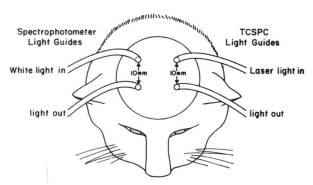

Fig. 12. Schematic of cat head showing light guide placement during hypoxia.

Blood Loss

This experiment studied the effect of the loss of blood on the intensity profiles. Readings were taken at 760 and 800 nm at varying levels of blood loss in an anesthetized dog. As with the hypoxia, the light fibers were placed about 10 mm apart on an exposed, but intact, skull. Readings were taken at normal FiO_2 (~ 21%) and blood samples were taken after each loss level was reached (and the dog had equilibrated) in order to determine hematocrit.

Figure 17 shows the intensity profiles vs. time for the four levels of blood volume lost (total blood volume ~ 600 ml). As blood loss increases, the tails of the logarithmic plots begin to lose their linearity (the tissue becomes more like a scatterer). In addition, the magnitude of the main slopes decrease corresponding to decreased absorption. This loss of absorption is plotted against hematocrit (% red blood cells in blood) in Figure 18. Since these curves are linear with respect to hematocrit (and therefore with respect to concentration), they show consistency with the Beer-Lambert Law. The extrapolation to 0% gives values different from zero indicating pure scattering of the animal tissue in the 0.05-0.06 cm^{-1} range.

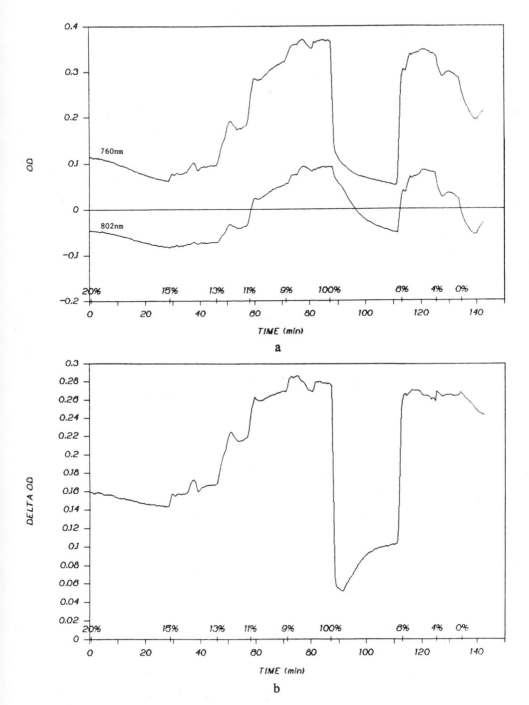

Fig. 13. Hypoxia data taken with scanning spectrophotometer: (a) 760 and 802 nm; (b) (760 - 802). The abscissa indicates inspired O_2.

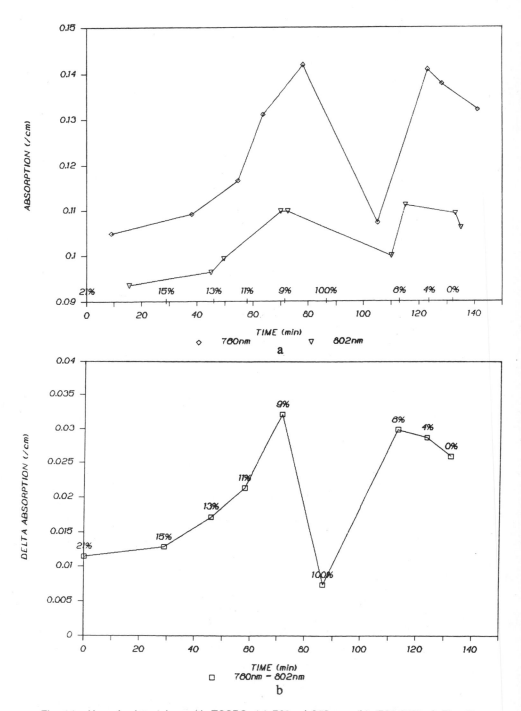

Fig. 14. Hypoxia data taken with TCSPC: (a) 760 nd 802 nm; (b) (760-802) cf. Fig. 13.

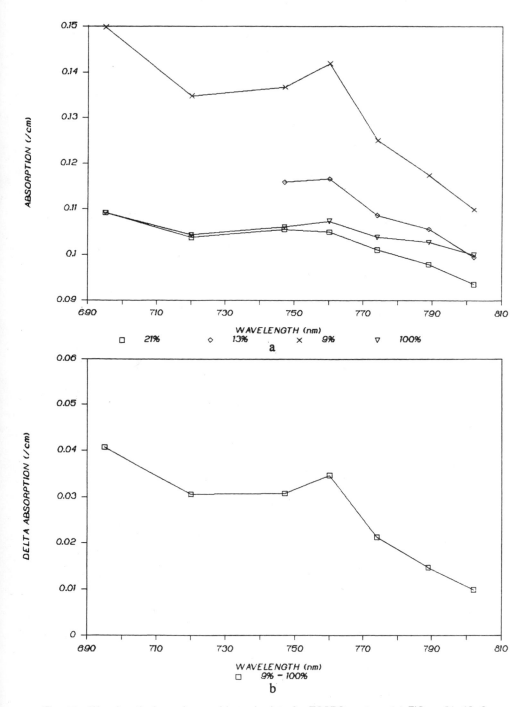

Fig. 15. Wavelength dependence of hypoxia data for TCSPC system: (a) FiO_2 = 21, 13, 9, 100%; (b) 9% - 100%.

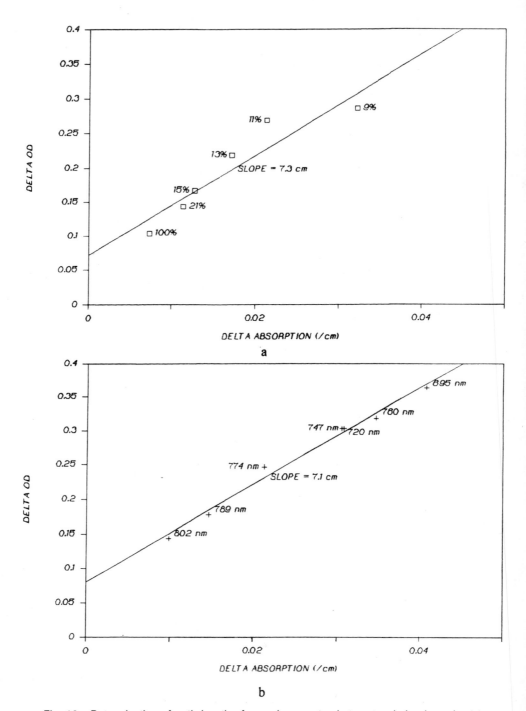

Fig. 16. Determination of path length of scanning spectrophotometer during hypoxia. (a) Parameter is FiO$_2$; (b) Parameter is λ.

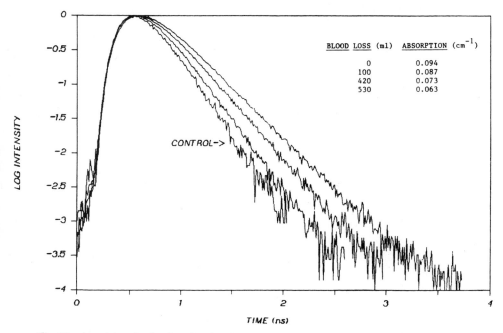

Fig 17. Log intensity for four levels of blood loss (increasing left to right) at 760 nm.

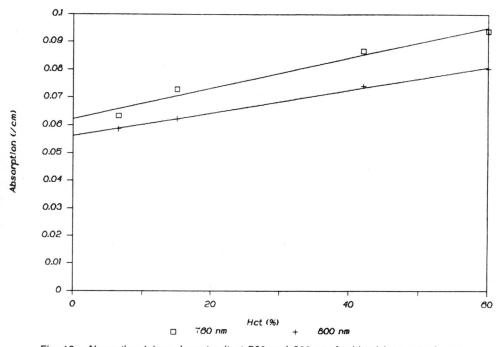

Fig. 18. Absorption (μ) vs. hematocrit at 760 and 800 nm for blood loss experiment.

COMPARISON OF MILK MODEL TO OTHER EXPERIMENTS

Now that we have a better idea of the effectiveness of the time-resolved system, we can evaluate the milk model. First, as was shown in Figure 17, with a decrease in blood, the tails of the log intensity become non-linear. Figure 19 demonstrates the similarity of the milk model with a 15 mm fiber separation to the lowest hematocrit curve (6.5%) for the blood loss experiment (7).

Comparing the milk model to the brain (Figure 6), we see in Figure 20 that for milk depths (no absorber) of 0 and 2.3 cm (early and late phases), there is a good trend as a function of fiber separation. If an absorber were added, the μ of the 2.3 cm curves would increase (see Figure 9). As with the initial time-resolved studies and the determination of characteristic lengths, this is an indication that we can see into the brain to a depth of several centimeters beyond the skull.

CONCLUSIONS AND FUTURE CONSIDERATIONS

This study of a model to mimic photon migration in tissue laid the background of optical theory in tissues by presenting absorption and scattering theories and discussing problems with them. Then it discussed how the theories led to earlier models of photon diffusion in scattering media and listed other physical models that have been used. By considering these other models along with optical properties of tissues, a homogeneous milk and blood model was developed and its attenuation curves as functions of fiber distance and milk depth were measured. Next, the theory of time-resolved spectroscopy was introduced to try and solve a great problem in tissue oximetry: finding out where the photons have been and what information they carry. By first using a high resolution system, we determined that absorption differences can be detected in a scattering medium. Next, we realized that time-correlation of the input light function is important in trying to determine total pathlength of the photons. In general, we are looking at the later arriving photons and, since we are taking the logarithm of the intensity, the amplitude of the input is irrelevant. Using animal models, we saw that the skull has certain effects that are very hard to model, but since the bulk of the received signal comes from the brain itself, the skull only has a minimal effect. In addition, even though the tissue should not follow the Beer Lambert Law (due to multiple scattering), we have shown that we get Beer-Lambert dependence that can be related to changes in absorption. Thus the major results shown here are:

- With a simple model it is possible to simulate photon migration in the brain;

- With a time-resolved system we can get total pathlength, absorption changes, and penetration depth;

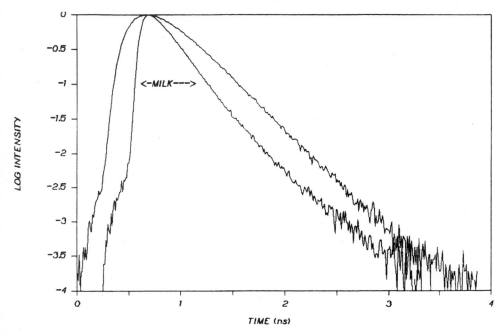

Fig. 19. Log intensity spectra of milk model (15 mm fiber separation) and low animal hematocrit.

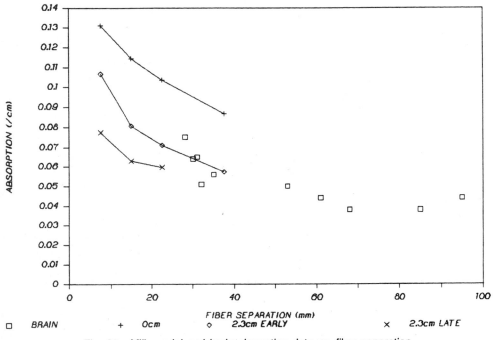

Fig. 20. Milk model and brain absorption data vs. fiber separation.

- Using both time-resolved and time-averaged systems, we can calibrate the pathlength of the latter;

- We may have to approach photon migration differently from before. It may be best to think of it less as travel through the brain from input to output and more as photons being introduced as a "bolus of light" which then acts as a point source emitting from the brain (hence the time shift). Also, it is possible that for large fiber separations, photons do not act according to a classical diffusion model.

Since we now know the model is effective in mimicking brain tissue in a simple way, it may be possible to modify it to model inhomogeneities within brain tissue (white vs. grey matter). This can lead to identification of abnormal or cancerous tissue which have different optical properties from normal tissue[9]. Also, by varying concentrations of scatterer and absorber, we should be able to mimic other tissues. Possibly taking into account more information about how tissues vary among people as well as what happens in extreme situations will lead to more universal applicability.

REFERENCES

1. B. Leskovar, C.C. Lo, P.R. Hartig and K. Sauer, Photon counting system for subnanosecond fluorescence lifetime measurements, *Rev. Sci. Instrum.*, 47:1113 (1976).
2. C. Lewis, W.R. Ware, L.J. Doemeny and T.L. Nemzek, The measurement of short-lived fluorescence decay using the single photon counting method, *Rev. Sci. Instrum.*, 44:107 (1973).
3. K.G. Spears, L.E. Cramer and L.D. Hoffland, Subnanosecond time-correlated photon counting with tunable lasers, *Rev. Sci. Instrum.*, 49:255 (1978).
4. M.C. Chang, S.H. Courtney, A.J. Cross, R.J. Gulotty, J.W. Petrich and G.R. Fleming, Time-correlated single photon counting with microchannel plate detectors, *Anal. Instrum.*, 14:433 (1985).
5. I. Yamazaki, N. Tamai, H. Kume, H. Tsuchiya and K. Oka, Micorchannel-plate photomultiplier applicability to the time-correlated photon-counting method, *Rev. Sci. Instrum.*, 57:1116 (1986).
6. R.F. Bonner, R. Nossal, S. Havlin and G.H. Weiss, Model for photon migration in turbid biological media, *J. Opt. Soc. Am. A*, 4(3):423 (1987).
7. B. Chance, J.S. Leigh, H. Miyake, D.S. Smith, S. Nioka, R. Greenfeld, M. Finander, K. Kaufmann, W. Levy, M. Young, P. Cohen, H. Yoshioka and R. Boretsky, Comparison of time resolved and unresolved measurements of deoxyhemoglobin in brain, *Proc. Natl. Acad. Sci. USA*, 1988. submitted 2/88.
8. B. Chance, S. Nioka, J. Kent, K. McCully, M. Fountain, R. Greenfeld and G. Holtom, Time resolved spectroscopy of hemoglobin and myoglobin in resting and ischemic muscle, *Anal. Biochem.*, 1988. submitted 4/88.
9. H.R. Eggert and V. Blazek, Optical properties of human brain tissue, meninges, and brain tumors in the spectral range of 200 to 900 nm, *Neurosurgery*, 21(4):459 (1987).

GIGAHERTZ FREQUENCY-DOMAIN FLUOROMETRY: RESOLUTION OF COMPLEX INTENSITY DECAYS, PICOSECOND PROCESSES AND FUTURE DEVELOPMENTS

Joseph R. Lakowicz

Department of Biological Chemistry
University of Maryland
School of Medicine
660 W. Redwood Street
Baltimore, Maryland   21201

SUMMARY

In this article we describe the principles, instrumentation and applications of frequency-domain fluorescence spectroscopy. This method can be used to evaluate complex time-dependent processes, even when the characteristic times are below 1 ns. This method is hence complementary to time-correlated single photon counting, in that one measures the frequency-response of the emission to modulated excitation, instead of the time response of the emission to pulsed excitation. To date, there have been relatively few attempts to measure photon migration in tissues by the frequency-domain method. We will demonstrate its potential for such measurements by demonstrating the resolution of multi-exponential decays on the psec timescale. The present instrumentation allows measurements to 2 GHz, which is easily adequate for measuring decay times as short as 50 ps. In the future it should be possible to extend the frequency range to 4 GHz, which should allow still faster processes to be quantified. Also, we have performed preliminary experiments in strongly scattering solutions which indicate that the phase shifts of the scattered light depend upon the amount of light absorption by the sample. It should be noted that this resolution of fast processes is not obtained at the expense of loosing information on the slower processes. Additionally, the GHz frequency-domain measurements are performed using low excitation intensities which do not damage the samples.

Photon Migration in Tissues
Edited by B. Chance
Plenum Press, New York

INTRODUCTION

Fluorescence spectroscopy is widely used to investigate the characteristics of biological macromolecules. Because of their complex photo-physical properties there have been considerable efforts to develop instrumentation which quantifies the ns and ps emission with high sensitivity, high signal-to-noise, and no systematic errors. Until recently, most measurements of time-dependent processes were accomplished by measurements in the time-domain[1-5].

During the past four years there has been remarkable progress in the alternative technique of frequency-domain fluorometry[6-8]. In this method one measures the frequency response of the sample to light which is intensity modulated over a wide range of frequencies, that is, the frequency-dependent phase angle and modulation. The early phase-modulation fluorometers operated at only one to three fixed frequencies[9-13], and these limited data were not adequate to recover multi-exponential decays of intensity or anisotropy[14]. During the intervening years several reports appeared describing phase-fluorometers which operated over a wide range of frequencies[15-17]. However, relatively little data has appeared from these instruments, and it is not clear that their precision is adequate to support the analysis of complex fluorescence decays. This situation has now changed, and instrumentation is commercially available which provides phase and modulation measurements from 2 to 200 MHz[18-19].

In general, the time and frequency-domain fluorometers have not been used to examine scattered light. This is because for most samples prepared for spectroscopic investigation the extent of scattering is small. Since there are relatively few scattering events there is no significant time-dependence in the scattered light. Hence, the scattered signal has only been useful as a zero-time reference signal in either the time or the frequency domains. However, the situation is different in highly scattering media such as tissue. In this case, the scattered photons have been subjected to multiple scattering events. The effective path length and time-dependence of the scattered light is dependent upon the nature of the tissue and the density of absorbing species. While little is currently known about the time-dependence of scattered light in tissue, it seems probable that this dependence will be multi- or non-exponential, as has been the case with the emission from complex molecules.

In this chapter we describe the principles and instrumentation of the frequency-domain method. In particular we will describe the first generation instrument, which operated to 200 MHz, and our state-of-the-art instrument which operates to 2 GHz. We overcame the 200 MHz limit using two modifications of the original design[8]. First, we avoided the use of light modulators by using the intrinsic high frequency harmonic content of a laser pulse train[20-21]. Secondly, we used a microchannel plate (MCP) photomultiplier (Hamamatsu R1564U). The single photoelectron pulse width of these devices (about 100 ps) is 10 to 20-fold less than that of a standard PMT[22,23]. Hence, we expected its bandwidth to extend to 2 GHz. To illustrate the potential usefulness of these instruments for studies of photon migration in tissues we describe the resolution of multi-exponential decays of fluorescence when the decay times range from 50 to 800 ps.

TIME AND FREQUENCY-DOMAIN FLUORESCENCE

The objective of most time-resolved measurements is to resolve the time-dependent emission from complex samples. The time-resolved emission is usually modeled as the sum of exponential decays. This model has gained widespread acceptance because it is simple and powerful enough to account for most measurements, and because of the intuitive link between the decay times and presumed components in the sample. However, it should be remembered that the actual decay may be non-exponential, i.e., not the sum of exponential components.

Suppose the sample contains two fluorescent substances, with decay times τ_1 and τ_2. Then, following a δ-function pulsed excitation the intensity decays as

$$I(t) = \sum_{i=0}^{2} \alpha_i e^{-t/\tau_i} \tag{1}$$

The fraction of the intensity observed in the usual steady state measurement due to each component is

$$f_i = \frac{\alpha_i \tau_i}{\sum_j \alpha_j \tau_j} \tag{2}$$

Time-resolved fluorescence data are most often obtained by direct measurements in the time-domain (Figure 1). The sample is excited with a pulse of light, and the emission detected with a detection system capable of nanosecond or picosecond time resolution. If the emission decays with a single decay time (Figure 1, solid line) it is rather easy to measure the decay time with good accuracy. The more difficult task is recovery of multiple decay times, which is illustrated for two widely spaced decay times in Figure 1 (0.2 and 1.0 ns, dashed line). It is generally difficult to resolve sums of exponentials because the parameters describing the decay are highly correlated. Hence, one requires a high signal-to-noise ratio, or equivalently a large number of photons, to recover the multiple decay times with reasonable confidence. Additionally, the pulse width of flash lamps is typically near 2 ns, which results in the need for extensive deconvolution of the data from overlap with the excitation pulse. Even if one uses a ps laser as the excitation source, the observed excitation function is often about 0.1 to 0.2 ns in width, due to timing jitter in the PMT and/or the timing electronics. Hence, it has been rather difficult to obtain data adequate to resolve multi-exponential decays, and therefore mixtures of fluorophores, especially when

171

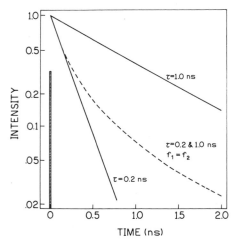

Fig. 1. Time-domain measurements of fluorescence decays. The sample is excited with a brief pulse of light. The solid lines shows the intensity decays for single decay times of 0.2 and 1.0 ns. The dashed line shows the decay for $\tau_1 = 0.2$ ns, $\tau_2 = 1.0$ ns, $f_1 = f_2 = 0.5$ (eqn. 2).

the decay times are below 1 ns. Resolution of complex or multi-exponential decays requires data with a high signal-to-noise ratio, which is difficult on the fluorescence timescale. We now show how a complex emission can be resolved in a novel way, from the frequency-response of the emission to a modulated light source.

The excitation source and measured values are different for frequency-domain fluorescence. The pulsed source is replaced with an intensity modulated light source. Because of the time lag between absorption and emission, the emission is delayed in time relative to the modulated excitation (Figure 2). The delay is described as the phase shift (Φ_ω). The finite time-response of the sample also results in demodulation of the emission by a factor m_ω.

Most samples of interest display more than one decay time. In these cases it is necessary to measure the phase and modulation values over the widest possible range of modulation frequencies, with the center frequency being comparable to the reciprocal of the mean decay time of the emission. The phase angle and modulation, measured over a wide range of frequencies, constitutes the frequency-response of the emission.

The shape of the frequency response is determined by the number of decay times displayed by the sample. If the decay is single exponential (Figure 3, top), the frequency-response is simple. If the sample displays two decay times (0.2 and 1.0 ns, Figure 3, bottom), a more complex response is observed. The objective is to recover the multiple

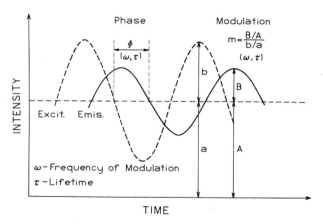

Fig. 2. Phase and modulation of fluorescence in response to intensity modulated excitation.

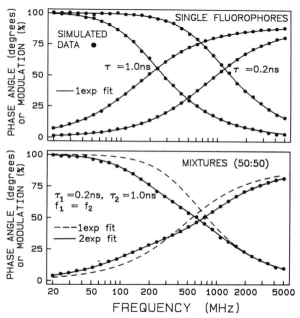

Fig. 3. Simulated frequency-domain data for single (top) and double (bottom) exponential decays. The phase angles increase and the modulation decreases with increasing modulation frequency. The dots (·) indicate the simulated data. Top: The solid lines show the best fits to a single decay time. Bottom: The dashed and solid lines show the best single and double exponential fits, respectively. The values of χ_R^2 are 1900. and 2.8 for the single and double exponential fits, respectively.

decay times from the experimentally measured frequency-response. This is generally accomplished by non-linear least-squares fitting of the measured phase and modulation values to various models[24-25].

The basis of all least-squares fitting procedures is comparison of the measured values with those predicted for various models. It is possible to predict the phase and modulation values for any assumed decay law, regardless of whether the detected light is due to emission or to scatter. The frequency-domain data can be calculated from the sine and cosine transforms of I(t)

$$N_\omega = \frac{\int_0^\infty I(t)\ \sin \omega t\ dt}{\int_0^\infty I(t)\ dt}$$

(3)

$$D_\omega = \frac{\int_0^\infty I(t)\ \cos \omega t\ dt}{\int_0^\infty I(t)\ dt}$$

(4)

where ω is the circular modulation frequency (2π x frequency in Hz). Even if the decay is more complex than a sum of exponentials it is generally adequate to approximate the decay by such a sum. If needed, nonexponential decay laws can be transformed numerically. For a sum of exponentials the transforms are

$$N_\omega = \sum_i \frac{\alpha_i\ \omega \tau_i^2}{(1 + \omega^2 \tau_i^2)} / \sum_i \alpha_i \tau_i$$

(5)

$$D_\omega = \sum_i \frac{\alpha_i \tau_i}{(1 + \omega^2 \tau_i^2)} / \sum_i \alpha_i \tau_i$$

(6)

The frequency-dependent values of the phase angle ($\Phi_{c\omega}$) and the demodulation ($m_{c\omega}$) are

$$\tan \Phi_{c\omega} = N_\omega / D_\omega$$

(7)

$$m_{c\omega} = \left[N_\omega^2 + D_\omega^2 \right]^{1/2}$$

(8)

The parameters (α_i and τ_i) are varied to yield the best fit between the data and the calculated values, as indicated by a miniumum value for the goodness-of-fit parameter χ_R^2,

$$\chi_R^2 = \frac{1}{\upsilon} \sum_\omega \left[\frac{\Phi_\omega - \Phi_{\omega c}}{\delta \Phi} \right]^2 + \frac{1}{\upsilon} \sum_\omega \left[\frac{m_\omega - m_{\omega c}}{\delta m} \right]^2$$

(9)

where υ is the number of degrees of freedom. The subscript c is used to indicate calculated values based upon the assumed parameters of the decay (α_i and τ_i), and $\delta \Phi$ and δm are the uncertainties in the phase and modulation values, respectively. The correctness of a model

174

is judged by the values of χ_R^2. For an appropriate model and random noise χ_R^2 is expected to be near unity. If χ_R^2 is greater than unity then one should consider whether the χ_R^2 value is adequate to reject the model. Rejection is judged from the probability that random noise could be the origin of the value of χ_R^2. This fitting procedure is illustrated by the solid and dashed lines in Figure 3. For the single exponential decays (top) it is possible to obtain a good match between the dots (·) and the curves calculated using the single exponential model. For a double exponential decay (bottom) the data (·) cannot be matched using a single decay time (---). However, the complex frequency-response is accounted for by the double exponential model (———) with the expected decay times (0.2 and 1 ns) and fractional intensities ($f_1 = f_2 = 0.5$).

INSTRUMENTATION FOR FREQUENCY-DOMAIN FLUOROMETRY

A schematic of our first-generation[7] frequency-domain fluorometer is shown in Figure 4. The instrument is similar to a standard fluorometer. The main differences are the laser light source, the modulator, and the associated radio frequency electronics. Until recently, it was thought to be difficult to obtain wideband light modulation. This is now known to be easily accomplished up to 200 MHz using standard electro-optic modulators[6,7]. In fact, the selection of a laser source in the first instruments was guided by the misconception that only lasers could be intensity modulated at adequate frequencies. It is presently possible to modulate arc lamps to 200 MHz[18,19], which will decrease the cost and complexity of the instrumentation. However, the strong collimation requirements and the large half-wave voltages of most electro-optic modulators results in rather low excitation intensities from the modulated arc lamp source.

At first glance it may appear difficult to measure the phase angles and modulation values at high frequencies. The measurements appear more difficult when one realizes that resolution of multi-exponential decays requires accuracy greater than 0.5° in phase and 0.5% in modulation. In fact, the measurements are surprisingly easy and free of interference. This is because of cross-correlation detection. The gain of the detector is modulated at a frequency offset ($F + \delta F$) from that of the modulated excitation (F). The difference frequency (δF) near 25 Hz contains the same phase and modulation information which would be observed directly at the actual modulation frequency (F). Hence, at all modulation frequencies, the phase and modulation can be measured with a simple zero-crossing detector and a ratio digital voltmeter. The use of cross-correlation detection results in the rejection of harmonics and other sources of noise. Consequently, there are no requirements that the excitation be a pure sinusoid. In fact, the frequency-components of almost any excitation profile can be used if it contains frequency components which are synchronized with the gain modulation of the detector.

Instrumentation of the type shown in Figure 4 has been used to recover the decay times of mixtures of fluorophores[25], decay time distributions[26,27], time-resolved emission spectra[28,29], analysis of excited state reactions[30], and resolution of associated and non-associated anisotropy decays[31-32].

GIGAHERTZ HARMONIC CONTENT FLUOROMETERS

The present configuration of our frequency-domain fluorometer is shown in Figure 5. The light source is a mode-locked argon ion laser. The ion laser pumps a dye laser, whose output is cavity dumped at 7.5862 MHz. The dye laser with R6G provides excitation wavelengths from 570 to 600 nm. For ultraviolet excitation the dye laser output is frequency doubled to 285-300 nm. For excitation from 355 to 380 nm we use the frequency-doubled and cavity-dumped output of a pyridine II dye laser.

Prior to installation of the pulsed laser source we used a frequency-doubled ring dye laser to obtain 280-305 nm excitation[33]. We found this source to be unstable and difficult to align and maintain. Indeed, procedures to align ring dye lasers have been the main subject of recent publications[34]. This continuous source required external modulation, which results in loss of about 90% of the light. The use of a continuous source also limited the modulation frequency to 200 MHz due to the limited bandwidth of light modulators with suitable UV transmission, thermal stability and optical geometry.

In comparison with a continuous laser source, a high repetition rate pulse laser provides several advantages. The high peak power of the ps pulses results in rather easy and

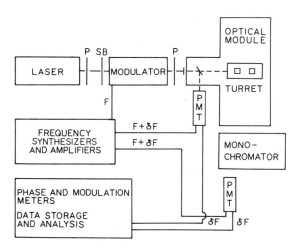

Fig. 4. Schematic of the variable-frequency phase-modulation fluorometer. P = polarizer, SB = Soleil-Babinet compensator, F = frequency, δF = cross-correlation frequency, and PMT = photo-multiplier tube.

efficient frequency-doubling. This UV output is essential for excitation of most fluorophores of interest, especially tyrosine and tryptophan fluorescence from proteins. The pulsed laser output is intrinsically modulated over a wide range of frequencies, as shown by the Fourier transform in Figure 5. Because of this intrinsic modulation the output can be used directly, rather than after passage through a modulator and associated optics, which results in a considerable loss of intensity and decreased sensitivity. Thirdly, the output of the pulser laser system was found to be considerably more stable and easier to align than that of our doubled ring dye laser.

An important feature of the pulsed laser source is that it contains harmonic output to many GHz, beyond that obtainable with any commercially available broadband modulator. A second interesting aspect of the harmonic method is that the Fourier components are greater than that possible with a 100% sine-modulated source[35]. Also, the dispersion of the harmonic content over a wide range of frequencies does not decrease the modulation available at each frequency. This is because all the photons in the light pulse contribute to the measurements at each frequency[36]. Basically, the detector integrates the complex emission with the desired cross-correlation frequency. Hence, the S/N ratio is expected to be comparable or superior to that obtainable using sinusoidal modulation.

Another important feature of the instrument is the use of a MCP PMT as the detector. As presently available, the MCP PMTs do not allow internal cross-correlation, which was used in the earlier frequency-domain instrument (Figure 4). We modified a 2 GHz

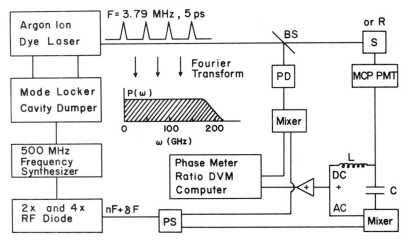

Fig. 5. 2 GHz frequency-domain fluorometer. PD - photodiode; PS - power splitter; MCP PMT - microchannel plate photomultiplier tube; BS - beam splitter; F - fundamental frequency of cavity dumped dye laser output; δF - cross correlation frequency of 25 Hz; N - number of the harmonic; S - sample; R - reference or scatter. From [15].

amplifier to allow detection and amplification of the DC component of the modulated intensity, which is essential for measurement of the frequency-dependent demodulation. The details are provided in reference 8.

The frequency range of the measurements is determined primarily by the modulation available at each frequency. The frequency-response of the entire system (laser source, amplifier and MCP-PMT) is shown in Figure 6. For phase modulation fluorometry a

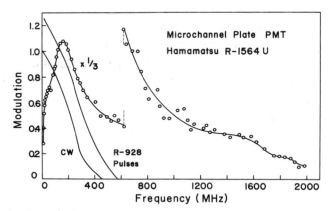

Fig. 6. Frequency-dependent modulation of scattered light. The open circles are the values found for the R1564-U MCP PMT. The other lines are the approximate modulation profiles for a R928, as found for a 5 ps pulse train or a modulated CW laser beam.

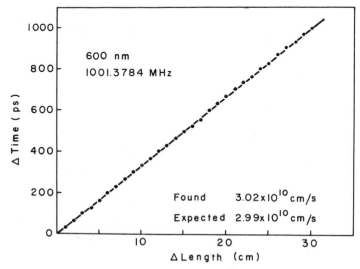

Fig. 7. Time-delay test using a variable distance. The time delays were calculated using t (ps) = $\Delta\Phi$ (deg)/freq (GHz) x 0.36°.

modulation in excess of 20% is adequate for most measurements. The 20% limit is near 2 GHz for the MCP PMT. Modulation in excess of 50% was found to 1 GHz.

The performance of the instrument was evaluated in several ways. The measured time delay was linear with distance, and the measured speed of light is within 1% of the expected value. The time delays were accurate to within ± 2 ps, with only seconds of data averaging for each distance (Figure 7). Our favorite test is the use of a calibrated 25 ps quartz plate from a Coherent auto-correlator (courtesy of Mr. Fred Gonzalez). The accuracy of the instrument was examined over a range of frequencies (Figure 8). The phase angles increased as expected and the measured time delays were within 2 psec of the expected value. On average, we found 27 ps for the presumed 25 ps delay. No demodulation is expected for a time delay, and no demodulation was observed to 2 GHz (Figure 8).

The ability of the GHz phase and modulation measurements to determine short decay times is illustrated in Figure 9, which shows the frequency-response of chromatographically purified rose bengal. The data (·) are in agreement with a single decay time model, with a decay time of 66 ps. Examination of Figure 9 reveals that we have not recovered the entire frequency response for this short lifetime. However, this particular MCP PMT (R2566) displays an exceptional bandwidth, and will permit future measurement to be extended to about 5 GHz (see Figure 13). In our present circuits the bandwidth is limited to 2 GHz because of the amplifiers and frequency synthesizers, not because of the PMT itself. The lower panels of Figure 9 show the deviations between the data and the single decay time model. These deviations are randomly distributed around zero, which indicates the absence of systematic errors throughout the entire 2 GHz bandwidth.

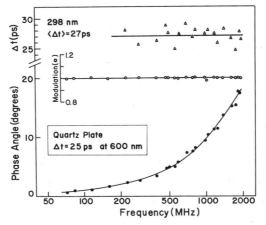

Fig. 8. Time-delay measurement using a 25-ps etalon quartz plate. This calibrated antireflection coated plate is from a Coherent autocorrelator, model 290. The wavelength was 298 nm.

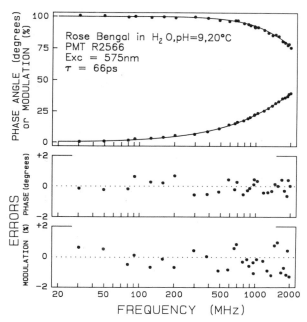

Fig. 9. Measurement of a picosecond decay time by harmonic content fluorometry. The sample was rose bengal in water. The rose bengal was purified chromatographically prior to the measurements.

RESOLUTION OF MULTI-EXPONENTIAL DECAYS

A common application of the frequency-domain measurements is the resolution of multi-exponential decays. To illustrate this application we examined a mixture of t-stilbene and p-quaterphenyl, which individually displayed decay times of 51 and 786 ps, respectively (Figure 10, top). The data ($\cdot$, o) measured for each compound individually are well matched by the curves calculated for the single decay time model (———). The lower panel (Figure 10) shows the frequency response of the mixture. The presence of multiple decay times is immediately evident from the more complex shape of the frequency response. The heterogeneity of the intensity decay is also evident from an attempt to fit the data to a single decay time ($\chi_R^2 = 1900$). The best fit, shown as the dashed line, is obviously inadequate. In contrast, the data can be explained by a curve with decay times of 50 and 756 ns (solid lines), $\chi_R^2 = 2.8$.

We note that the resolution shown in Fig. 10 is a simple one in that the decay times are different by 15-fold. In other studies we have demonstrated that more closely spaced lifetimes can be resolved, as can the decay times of three component mixtures. On the other hand the mixture of t-stilbene and p-quaterphenyl is difficult because both decay times are less than 1 ns, and the decay time of t-stilbene is only 50 ps.

180

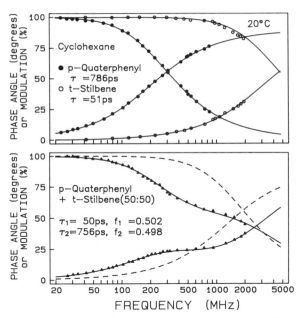

Fig. 10. Resolution of a double exponential decay. The sample consisted of a mixture of t-stilbene
and p-quaterphenyl which individually displayed decay times of 51 and 786 ps,
respectively.

EXTENSION BEYOND 2 GHz

Examination of Figure 9 revealed that the 2 GHz measurements were not adequate to
recover the entire frequency-response of the sample. Of course, resolution of complex
decays and confidence in the resulting parameters is enhanced by data which span the entire
frequency response of the emission.

What are the possibilities of frequency-domain measurements at frequencies above 2
GHz? Two years ago such measurements did not seem possible, as the R1564U was the
fastest available MCP PMT. However, this situation has changed as the result of the
introduction of two faster PMTs, these being the R2566 with either 6 or 12 micron
channels (Hamamatsu, Inc.). These PMTs have a grid between the second MCP and the anode,
which decreases the pulse width due to a single photoelectron. The main effect of the anode
is on the fall time of the pulse. Since the frequency-response of a PMT depends on the total
pulse width, this can be important for frequency-domain measurements. However, the
R2566 MCP PMTs may not have any advantage for time-correlated single photon counting,
since these measurements rely on the rise time of the pulses.

We measured the frequency response of several detectors, the R928, R1564U,
R2566-12, R2566-6, and a 25 ps photodiode (Figure 11). The improved frequency range
of the R2566-12, relative to the R1564U, is immediately apparent. The R1564U is known

provides data to 2 GHz, with no detectable increase in our S/N ratio. By analogy, the R2566-12 will allow measurements at least 4.3 GHz (Figure 11). Subsequently, a R2566 was fabricated with 6 micron microchannel plates. To date we have not had an opportunity to test this device. However, the manufacturer has provided the calculated frequency response, which we redrew in Figure 12. Using the 10% level as a cut-off, the frequency-domain data will easily be measurable to 6 GHz. We are optimistic that data can be obtained to the 1% cutoffs, which means that 10 GHz measurements should be possible within the foreseeable future.

PHOTON MIGRATION IN A SCATTERING MEDIUM

And finally, we examined the phase shifts due to light scattered through a turbid medium (Figure 13). We selected a non-dairy coffee creamer as the scatterer, and

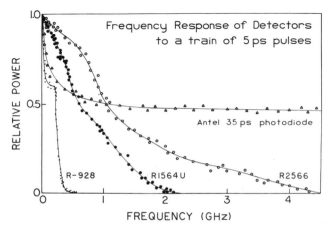

Fig. 11. Measured frequency-response of several PMTs, and a fast photodiode. These data were obtained using an Anritsu Spectrum Analyzer, which was kindly loaned to us by Wayne Zimmerman, Creative Marketing, Rockville, Maryland.

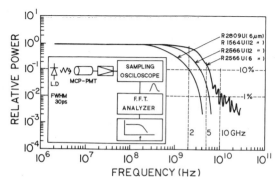

Fig. 12. Frequency-response of microchannel plate PMTs. Redrawn from technical brochures distributed by Hamamatsu, Inc.

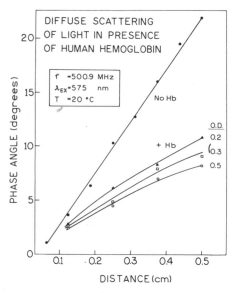

Fig. 13. Measurement of light diffusion by the frequency-domain method.

hemoglobin was added as the absorber. The input and output signals were through fiber optics. The phase shifts were found to depend upon the distance between the fiber tips. More importantly, the phase angles were dependent upon the amount of hemoglobin added to the scatter. This dependence of phase angle on absorption, and the well-known dependence of hemoglobin and cytochrome absorbance of the oxygenation state of tissues, suggests that the phase angles could be used for in-vivo measurements of tissue oxygenation. Clearly, this topic must be explored in greater detail in future experiments.

ACKNOWLEDGMENTS

This work was supported by grants from the National Science Foundation (DMB-8502835 and 8511065) and from the National Institutes of Health (GM-35154 and GM-39617). The author thanks Dr. Henryk Szmacinski for providing the phase data for light scatter.

REFERENCES

1. R. R. Alfono, "Biological Events Probed by Ultrafast Laser Spectroscopy," Academic Press, New York (1982).
2. P. M. Bayley and R. E. Dale, "Spectroscopy and the Dynamics of Molecular Biological Systems," Academic Press, New York (1986).

3. R. B. Cundall and R. E. Dale, "Time-Resolved Fluorescence Spectroscopy in Biochemistry and Biology," Plenum NATO ASI Series, New York (1980).

4. J. N. Demas, "Excited State Lifetime Measurements," Academic Press, New York (1983).

5. D. V. O'Connor and D. Phillips, "Time-Correlated Single Photon Counting," Academic Press, New York (1984).

6. E. Gratton and M. Limkeman, A continuous variable frequency cross-correlation phase fluorometer with picosecond resolution, *Biophys. J.*, 44:315-324 (1983).

7. J. R. Lakowicz and B. P. Maliwal, Construction and performance of a variable frequency phase-modulation fluorometer, *Biophysical Chemistry*, 21:61-78 (1985).

8. J. R. Lakowicz, G. Laczko, and I. Gryczynski, A 2 GHz frequency-domain fluorometer, *Rev. Sci. Instrum.*, 57:2499-2506 (1986).

9. E. A. Bailey and G. K. Rollefson, The determination of the fluorescence lifetimes of dissolved substances by a phase shift method, *J. Chem. Phys.*, 21:1315-1326 (1953).

10. A. Muller, R. Lumry and H. Kokubun, High-performance phase fluorometer constructed from commercial subunits, *Rev. Sci. Instrum.*, 36:1214-1226 (1965).

11. A. M. Bonch-Breuvich, I. M. Kazarin, V. A. Molchanov, and I. V. Shirokov, An experimental model of a phase fluorometer, *Instrum. Exp. Tech. (USSR)*, 2:231-236 (1959).

12. R. D. Spencer and G. Weber, Measurement of subnanosecond fluorescence lifetimes with a cross-correlation phase fluorometer, *Ann. N.Y. Acad. Sci.*, 158:361-376 (1969).

13. U. K. A. Klein, Picosecond fluorescence decay studied by phasefluorometry of rotational diffusion in liquids, *The Arabian Journal for Sci. and Eng.*, 9:327-344 (1984).

14. D. M. Jameson and E. Gratton, Analysis of heterogeneous emissions by multifrequency phase and modulation fluorometry, in: "New directions in molecular luminescence, Amer. Soc. for Testing and Materials," D. Eastwood, ed., ASTM, Baltimore (1983).

15. M. Hauser and G. Heidt, Phase fluorometer with a continuously variable frequency, *Rev. Sci. Instrum.*, 46:470-471 (1975).

16. H. P. Haar and M. Hauser, Phase fluorometer for measurement of picosecond processes, *Rev. Sci. Inst.*, 49:632-633 (1978).

17. G. Ide, Y. Engelborghs, and A. Persoons, Fluorescence lifetime resolution with phase fluorometry, *Rev. Sci. Inst.*, 54:841-844 (1983).

18. SLM Instruments, Urbana, Illinois.

19. ISS Instruments, Urbana, Illinois.

20. E. Gratton and R. Lopez-Delgado, Measuring fluorescence decay times by phase-shift and modulation techniques using the high harmonic content of pulsed light sources, *Nuovo Cimento*, B56:110-124 (1980).

21. E. Gratton, D. M. Jameson, N. Rosato and G. Weber, Multifrequency cross-correlation phase fluorometer using synchroton radiation, *Rev. Sci. Inst.*, 55:486-494 (1984).

22. I. Yamazaki, N. Tami, H. Kume, H. Tsuchiya, and K. Oba, Microchannel-plate photomultiplier applicability to the time-correlated photon-counting method, *Rev. Sci. Inst.*, 56:1187-1194 (1985).

23. S. Kinoshita and T. Kushida, Picosecond fluorescence spectroscopy by time-correlated single-photon counting, *Analytical Instrumentation*, 14:503-524 (1985).

24. J. R. Lakowicz, E. Gratton, G. Laczko, H. Cherek and M. Limkeman, Analysis of fluorescence decay kinetics from variable-frequency phase shift and modulation data, *Biophysical Journal*, 46:463-477 (1984).

25. E. Gratton, J. R. Lakowicz, B. Maliwal, H. Cherek, G. Laczko and M. Limkeman, Resolution of mixtures of fluorophores using variable-frequency phase and modulation data, *Biophysical Journal*, 46:479-486 (1984).

26. E. Gratton, J. R. Alcala and F. G. Prendergast, Interpretation of fluorescence decays in proteins using continuous lifetime distributions, *Biophys. J. Biophysical Society*, 51:925-936 (1987).

27. J. R. Lakowicz, H. Cherek, I. Gryczynski, N. Joshi and M. L. Johnson, Analysis of fluorescence decay kinetics measured in the frequency-domain using distribution of decay times, *Biophysical Chemistry*, 28:35-50 (1987).

28. J. R. Lakowicz, H. Cherek, B. Maliwal, G. Laczko and E. Gratton, Determination of time-resolved fluorescence emission spectra and anisotropies of a fluorophore-protein complex using frequency-domain phase-modulation fluorometry, *J. Biol. Chem.*, 259:10967-10972 (1984).

29. J. R. Lakowicz, H. Cherek, G. Laczko and E. Gratton, Time-resolved fluorescence emission spectra of labeled phospholipid vesicles, as observed using frequency-domain fluorometry, *Biochim. Biophys. Acta*, 777:183-193 (1984).

30. J. R. Lakowicz and H. Cherek, Resolution of an excited state reaction using frequency-domain fluorometry, *Chem. Phys. Letters*, 122:380-384 (1986).

31. J. R. Lakowicz, H. Cherek, B. Maliwal and E. Gratton, Time-resolved fluorescence anisotropies of fluorophores in solvents and lipid bilayers obtained from frequency-domain phase-modulation fluorometry, *Biochemistry*, 24:376-383 (1985).

32. H. Szmacinski, R. Jayaweera, H. Cherek and J. R. Lakowicz, Demonstration of an associated anisotropy decay by frequency-domain fluorometry, *Biophysical Chemistry*, 27:233-241 (1987).

33. J.R. Lakowicz, I. Gryczynski and H. Cherek, Measurement of subnano-second anisotropy decays of protein fluorescence using frequency-domain fluorometry, *J. Biol. Chem.*, 261:2240-2245 (1986).

34. A.M. der Spek, P. Magendans, and A.H. Kemper, Improved cavity alignment system for spectra physics 380 D single-mode ring dye lasers, *Rev. Sci. Instrum.*, 59:653-654 (1988).

35. K. Berndt, H. Durr and D. Palmer, Picosecond phase fluorometry by mode-locked CW lasers, *Optics Communications*, 42:419-422 (1982).

36. J. R. Alcala, E. Gratton and D. M. Jameson, A multifrequency phase fluorometer using the harmonic content of a mode-locked laser, *Analytical Instrumentation*, 14:225-250 (1985).

Appendix

AUTHORS

R.R. Alfano
Institute for Ultrafast Spectroscopy and Lasers
Photonics Application Laboratory
Department of Electrical Engineering
The City College of New York
New York, NY 10031

Clyde H. Barlow
The Evergreen State College,
Olympia, Washington 98505 and
Department of Chemistry, University of Washington
Seattle, Washington 98195

David H. Burns
Center for Bioengineering, University of Washington
Seattle, Washington 98195

Robert F. Bonner
Biomedical Engineering and Instrumentation Branch
National Institutes of Health
Bethesda, MD 20892

P. Baldeck
Institute for Ultrafast Spectroscopy and Lasers
Photonics Application Laboratory
Department of Electrical Engineering
The City College of New York
New York, NY 10031

Benjamin J. Comfort
Department of Medicine
Duke University Medical Center
Durham, North Carolina 27710

James B. Callis

Department of Chemistry, University of Washington
Seattle, Washington 98195

B. Chance

Department of Biochemistry and Biophysics
University of Pennsylvania
Philadelphia, PA 19104 USA

Stephen T. Flock

Hamilton Regional Cancer Centre
and McMaster University
711 Concession Street
Hamilton, Ontario, Canada L8V 1C3

T. Hayakawa

Hamamatsu Photonics K.K.
1126-1 Ichino-cho, Hamamatsu City, Japan 435

P.P. Ho

Institute for Ultrafast Spectroscopy and Lasers
Photonics Application Laboratory
Department of Electrical Engineering
The City College of New York
New York, NY 10031

Gary R. Holtom

Regional Laser and Biotechnology Laboratories
University of Pennsylvania
Philadelphia, PA 19104 USA

Jay R. Knutson

Laboratory of Technical Development,
National Heart, Lung and Blood Institute
Building 10, Room 5D-10
Bethesda, MD 20892

Joseph R. Lakowicz

Department of Biological Chemistry
University of Maryland
School of Medicine
660 W. Redwood Street
Baltimore, Maryland 21201

M. Maris

Department of Biochemistry and Biophysics
University of Pennsylvania
Philadelphia, PA 19104 USA

H. Miyake

    Department of Biochemistry and Biophysics

    University of Pennsylvania

    Philadelphia, PA 19104 USA

S. Miyaki

    Hamamatsu Photonics K.K.

    1126-1 Ichino-cho, Hamamatsu City, Japan 435

Shoko Nioka

    Dept. of Biochemistry/Biophysics

    University of Pennsylvania

    Philadelphia, PA

Ralph Nossal

    Physical Sciences Laboratory

    National Institutes of Health

    Bethesda, MD 20892

Kouich Oka

    Otsuka Electronics, Ltd.

    Osaka, Japan

Michael S. Patterson

    Hamilton Regional Cancer Centre

    and McMaster University

    711 Concession Street

    Hamilton, Ontario, Canada L8V 1C3

Claude A. Piantadosi

    Department of Medicine

    Duke University Medical Center

    Durham, North Carolina 27710

F. Raccah

    Institute for Ultrafast Spectroscopy and Lasers

    Photonics Application Laboratory, Department of Electrical Engineering

    The City College of New York

    New York, NY 10031

D.S. Smith

    Department of Anesthesiology

    Hospital of the University of Pennsylvania

    Philadelphia, PA 19104

S. Suzuki

    Hamamatsu Photonics K.K.

    1126-1 Ichino-cho, Hamamatsu City, Japan 435

G. Tang

Institute for Ultrafast Spectroscopy and Lasers
Photonics Application Laboratory
Department of Electrical Engineering
The City College of New York
New York, NY 10031

George H. Weiss

Physical Sciences Laboratory
National Institutes of Health
Bethesda, MD 20892

Brian C. Wilson

Hamilton Regional Cancer Centre
and McMaster University
711 Concession Street
Hamilton, Ontario, Canada L8V 1C3

K.S. Wong

Institute for Ultrafast Spectroscopy and Lasers
Photonics Application Laboratory
Department of Electrical Engineering
The City College of New York
New York, NY 10031

INDEX

DATE DUE

FEB 2 7 1991			
NOV 0 4 1999			
AUG 2 0 2003			